Bacteria and Mineral Cycling

Bacteria
and
Mineral Cycling

T. Fenchel *and* T. H. Blackburn

Institute of Ecology and Genetics
Aarhus, Denmark

1979

ACADEMIC PRESS
LONDON   NEW YORK   SAN FRANCISCO

A Subsidiary of Harcourt Brace Jovanovich, Publishers

ACADEMIC PRESS INC. (LONDON) LTD.
24/28 Oval Road
London NW1

United States Edition published by
ACADEMIC PRESS INC.
111 Fifth Avenue
New York, New York 10003

British Library Cataloguing in Publication Data

Fenchel, T.
  Bacteria and mineral cycling.
  1. Bacteria—Physiology
  2. Microbial metabolism
  3. Microbial ecology
  I. Title   II. Blackburn T. Henry
  589.9′01′33   QR88   78–54532

  ISBN 0-12-252750-X

Typeset in Great Britain by
Kelmscott Press Ltd., 30 New Bridge Street, London EC4V 6BJ

Printed in Great Britain by
John Wright & Sons, Ltd., at the Stonebridge Press, Bristol

Preface

This book is a general text, describing the influences of bacterial metabolism on the chemical environment of the biosphere. Being a general review, it should be of value to those working in various aspects of this field. General microbiologists should find it interesting to read about the ecological significance of their favorite organisms, and ecologists lacking a background in microbiology should find this book useful for understanding the role of bacteria in natural ecosystems. This book will perhaps also prove useful as a textbook for graduate courses in microbial ecology. Finally, those working in more applied fields involving microbial ecosystems (e.g., soil science, environmental engineering) hopefully will find that our book may add a new and more general perspective to their fields of study.

Although our book presumes some elementary knowledge of general microbiology, biological energetics, and ecological principles, we include two appendices, as aids to the reader. The first appendix shows how free energy yields of biological processes are calculated. Appendix II is a simplified systematic index of the prokaryotes, with notes on their physiology and natural history. We have included a large number of references to recent literature, and we have reviewed work which has not previously found its way into textbooks.

Many recent books on microbiology and microbial ecology use rather fuzzy definitions of "microorganisms", often including, in addition to the bacteria, some part of the unicellular eukaryotes, up to the fungi, including mushrooms. Our book deals exclusively with prokaryotes, including the blue-green algae (referred to as blue-green bacteria or cyanobacteria). Research carried out in the last decade has shown that from structural, physiological, and evolutionary viewpoints, the distinction between prokaryotes and eukaryotes is the most fundamental classification of living organisms. This is, to a large extent, also true from an ecological point of view (discussed in detail in Chapters 1 and 3). We have, therefore, confined our discussion to prokaryote metabolism and ecology, although the importance of bacteria relative to eukaryotes with respect to carrying out different processes in natural ecosystems is discussed in several chapters.

Three aspects of our subject are stressed. The energetics of bacterial processes is treated in some detail and is used as a general framework for most of the topics of the book. Energetic considerations may yield much insight into the understanding of which processes are actually important in natural ecosystems. Secondly, evolutionary aspects are emphasized, viz., the chemical evolution of the biosphere which largely parallels prokaryote evolution. Finally, we have attempted to show how the different element cycles interconnect in natural ecosystems.

The first chapter is devoted to bacterial transformations of matter, as described in terms of energetics of the processes. In particular, we discuss in detail the dissimilatory reduction–oxidation processes involving carbon, nitrogen and sulfur, and the assimilatory reductions of these elements. This chapter gives a general background for the remaining portion of the book.

Chapter 2 discusses theories about the evolution of prokaryote metabolic pathways, in context with palaeontological and geochemical evidence and with current ideas on the evolution of the atmosphere. It is concluded that the chemical composition of the biosphere and its important biologically mediated element cycles are largely the product of prokaryote evolution.

The remainder of the book concentrates on the extant biosphere. In Chapter 3, it is shown that the microbial mineral cycling is mainly driven by the chemical energy of detritus, viz., dead organic material, which is mostly derived from plants. The dominating role of bacteria in the mineralization of this material, which in many ecosystems represents the dominant part of the primary productivity, is explored. These considerations are followed by a treatment of the mineralization of dissolved and particulate organic material in different types of habitats, the hydrolysis of structural compounds, the breakdown of hydrocarbons, and the microbial role in the fossilization of organic matter.

The following three chapters (4–6) are central to the book; they treat the microbial transformations of carbon, nitrogen, and sulfur compounds, respectively. Examples from several different types of habitats (sediments, soils, sewage treatment plants, etc.) are given, but first of all, the general principles of these element cycles and their interconnections are stressed.

Chapter 7 discusses bacterial transformations of other elements. Some attention is given to the controversial dissimilatory oxidation of iron and manganese compounds. This is followed by a short account of some bacterial transformations of trace metals and of the biological phosphorus cycle.

In Chapter 8, examples of mineral cycling connected with microbial symbiosis are discussed. Most emphasis is given to the symbiotic degradation of structural carbohydrates, especially in ruminants. This system is probably still the best understood natural microbial community, and it gives a good illustration of many of the principles discussed in previous chapters, particularly with respect to the ecology of fermentation. Symbiotic nitrogen fixation is also discussed at some length. Finally, the chapter offers a general discussion of mutualistic relationships between microorganisms and the theory of "serial endosymbiosis" for explaining the origin of the eukaryotic cell from the viewpoint of mineral cycling in symbiotic relationships.

We conclude with a chapter on the global cycling of carbon, nitrogen and sulfur, with special emphasis on the quantitative and qualitative role of prokaryotes. There are still great gaps in our knowledge with respect to causal relationships between processes, and in defining and quantifying transfer rates; this has often not been sufficiently emphasized by textbook authors or by prophets predicting the effects of human activity on the future of the biosphere. The chapter also serves as a summary of the most important principles of mineral cycling discussed in previous chapters.

The book covers a large field relative to the combined knowledge and research experiences of the authors. Readers may therefore, quite justifiably find that it is biased in some places, and that some topics and examples are over-emphasized whereas others are treated superficially, or even contain misunderstandings. Still, we hope that the general principles of the subject will be understood.

Acknowledgements. Our gratitude is due to E. Broda, J. E. Hobbie, and A. Oren who all read larger or smaller parts of the manuscript and contributed useful suggestions and corrections. A special gratitude is due to B. B. Jørgensen for his careful review of the whole manuscript and for his many constructive suggestions. Our thanks are also extended to Sheila Blackburn and Hilary Adler-Fenchel for linguistic improvements and for assistance in proof-reading and in the preparation of the subject index. Finally we acknowledge the Danish Science Research Council for its support over many years to our research group in microbial ecology.

Contents

CHAPTER 1

The Requirements of the Bacterial Cell

1.1. ENERGY YIELDING PROCESSES

Bacteria require a nutrient for only one of two processes (1) to utilize in energy yielding processes or (2) to assimilate for cell synthesis. These processes are closely related since a considerable portion of a cell's energy budget is expended on biosynthetic reactions (Table I). Protein biosynthesis accounts for about 60 % and nutrient transport for about 18 %. If the nutrients require to be reduced before assimilation into cell constituents, energy must be expended on the reductive processes. This is considered in detail in the second half of this chapter.

It is well established that adenosine triphosphate (ATP) is the main energy-coupling agent in all cells (Lehninger, 1971). It is generated in the reaction:

$$\underset{\substack{\text{adenosine}\\ \text{diphosphate}}}{\text{ADP}} + \underset{\text{phosphate}}{\text{Pi}} + \text{energy} \rightleftharpoons \underset{\substack{\text{adenosine}\\ \text{triphosphate}}}{\text{ATP}} + H_2O$$

When a chemical reaction results in the production of heat it is said to be *exothermic* and to have a negative change in *enthalpy* (ΔH is negative); the reaction proceeds in the forward direction, e.g.:

$$\text{glucose} + 6O_2 \rightleftharpoons 6CO_2 + 6H_2O; \qquad \Delta H = -673 \text{ kcal mol}^{-1}.$$

TABLE I. Bacterial Energy Budget.

Process	% Energy (ATP) expended on each process[a]
Synthesis:	
Polysaccharide	6·5
Protein	61·1
Lipid	0·4
Nucleic acid	13·5
Transport into cells	18·3

[a] Based on Stouthamer (1973) for a cell grown on glucose.

1

The maximum amount of energy that can be utilized in such a reaction is known as the free energy and is related to enthalpy change as follows:

$$\Delta G = \Delta H - T\Delta S,$$

when ΔG is the change in the free energy in the system, ΔH the heat transferred between the system and its surroundings, T the absolute temperature, and ΔS the entropy change in the system. The free energy can be calculated from the equilibrium constant of a reversible reaction:

$$\Delta G^\circ = - RT \ln K_{eq},$$

where R is the gas constant and K_{eq} is the equilibrium constant. The symbol ΔG° represents the change in *standard free energy*, when one mol of a reactant is converted to one mol of product at $25^\circ C$, 1 atm. The standard free energy at pH 7 is represented by $\Delta G'_o$. The standard free energy change, $\Delta G'_o$, for the oxidation of glucose is -686 kcal mol^{-1} and represents the maximum amount of energy that may be gained from the reaction. It is common practice to calculate ΔG° from the *energy of formation* of compounds in a reaction, as illustrated in Appendix I.

Another useful relationship in considering energy transformations in oxidation/reduction reactions is:

$$\Delta G^\circ = - nF \Delta\varepsilon,$$

where n is the number of electrons, F is the Faraday (23 kcal mol^{-1}) and $\Delta\varepsilon$ is the difference in standard oxidation/reduction potentials of the reactants.

The $\Delta G'_o$ for the hydrolysis of ATP is -7 kcal mol^{-1}:

$$ATP^{4-} + H_2O \rightleftharpoons ADP^{3-} + HPO_4^{2-} + H^+.$$

It therefore follows that at least 7 kcal mol^{-1} reactant must be available to generate one mol ATP. Reactions with a $-\Delta G'_o$ of less than 7 kcal mol^{-1} cannot be coupled directly to ATP generation although there are possibilities for indirect coupling which will be discussed later.

It was noted that $\Delta G'_o$ is defined using *molar* concentrations of reactants but since these concentrations are definitely non-biological, the calculated free energy changes are not always very precise. Table II shows a variety of energy yielding reactions which sustain bacterial growth, presumably because they are capable of generating ATP. Two principal methods of ATP synthesis are used by bacteria in these reactions: *substrate level phosphorylation* and *electron transport (oxidative) phosphorylation* of ADP. These are briefly described before the chemotrophic energy yielding reactions (Fig. 1) themselves are considered.

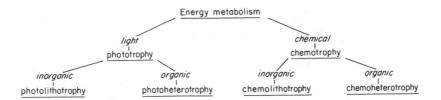

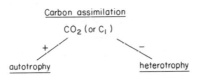

Fig. 1. Bacterial classification based on nutritional requirements. Above in terms of energy requirements and below in terms of carbon assimilation. An additional term, mixotroph, describes a facultative autotroph growing under conditions where it simultaneously uses autotrophic and heterotrophic pathways.

TABLE II. Chemotrophic Reactions Sustaining Bacterial Growth[a].

Oxidant (e^- acceptor)	Reductant (e^- donor)							
	H_2	CHO	CH_4	HS^-	NH_4^+	N_2	NO_2^-	Fe^{2+}
H^+/H_2O	−	+	−	−	−	−	−	−
CHO	+	+	−	−	−	−	−	−
CO_2	+	+	−	−	−	−	−	−
SO_4^{2-}	+	+	?	−	−	−	−	−
NO_3^-	+	+	?	+	−	−	−	−
O_2	+	+	+	+	+	−	+	+

[a] The electron donors are listed in order (left to right) of decreasing capacity to donate electrons (H, C, S, N, Fe). The electron acceptors are listed in order (top to bottom) of increasing capacity to accept electrons (H, C, S, N, O). There is thus an increasing difference in E'_o between the couples in going down a series and a decreasing difference in E'_o in going across a series. The larger the difference in E'_o values, the greater is the free energy that is available for the reaction. CHO is used to represent reduced carbon of undefined composition.

1. *Substrate level phosphorylation.* In the reactions:

$$\text{glyceraldehyde} + \text{phosphate} \xrightarrow[\text{(minus } 2e^-)]{\text{oxidation}} 2H + H_2O + \text{3-phosphoglyceric acid},$$

$$\text{3-phosphoglyceric acid} + ADP \rightarrow \text{glyceric acid} + ATP,$$

the $\Delta G'_o$ of -7 kcal for the oxidation of the aldehyde to the acid is preserved in ATP and the coupled reactions occur with a much smaller decrease in free energy. This is substrate level phosphorylation because one molecule of substrate must be phosphorylated for each molecule of ATP synthesized.

2. *Electron transport phosphorylation.* In this process no direct phosphorylation of a large quantity of compound occurs in the cells. The mechanism has not been completely elucidated but it may involve a reversal of ATPase, driven by an electrochemical proton gradient across a membrane containing the electron transporting components of a chain (Mitchell, 1977). The capacity of electrons to generate enough energy to synthesize ATP may be related to the change in potential.

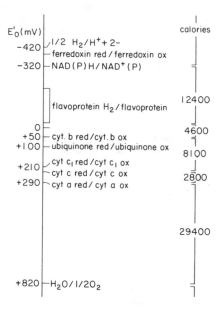

Fig. 2. An electron transport chain. On the left are shown the oxidation/reduction potentials of various couples in a representative electron transport chain. The distance between potentials reflects the energy difference between the corresponding electron carriers. This energy may be expressed in calories (right hand axis) where it is seen that sufficient energy is available between three carriers (12·4, 8·1 and 29·4 kcal) to generate ATP (7 kcal), if some form of direct coupling is necessary.

Standard oxidation/reduction potentials, E'_o, are based on the assignment of an arbitrary value of zero volts to the hydrogen electrode at pH 0 when H_2 at 1 atm is 50% ionized:

$$\tfrac{1}{2}H_2 \rightleftharpoons H^+ + e^-.$$

E'_o is the potential of the half reduced system at pH 7; it is -420 mV. At the other end of the scale we have:

$$\tfrac{1}{2}O_2 + 2e^- \rightleftharpoons O^{2-}; \qquad E'_o = +820 \text{ mV}.$$

Figure 2 shows the E'_o values for these two extremes (O_2 and H_2) with the intermediate values of the components of the mitochondrial or aerobic bacterial electron transporting chain. Many bacteria have cytochromes other than those shown in Fig. 2 (Jones, 1977). The $\Delta G'_o$ values between the components are given for comparison. A potential difference of 200 mV for a two electron transfer between donor and acceptor is necessary to give a $\Delta G'_o$ of -9 kcal mol^{-1}, or approximately the amount of energy required to ensure the generation of one ATP. This may not be necessary if a proton gradient drives phosphorylations. There are only three coupled reactions in the transport chain at which ATP is traditionally thought to be generated. The reduced state of the electron carrier NAD(P)H is generated partly during glycolysis but principally from the oxidation of acetate to carbon dioxide through the tricarboxylic acid (TCA) cycle; Lehninger (1971) should be consulted for further details. The glycolytic oxidation of glucose yields only 2 mol ATP per mol of substrate, whereas the electron transport phosphorylating system should yield a further 36 mol per mol substrate. Considerably more energy is thus obtained when the E'_o of the reactants are widely separated (CHO/O_2) than when they are close together (CHO/CHO) as in glycolysis (CHO is used to denote reduced organic carbon of undefined composition). It is more relevant to consider differences in E'_o between reacting compounds in defining energy yield than it is to define, e.g., glucose as a "high energy" nutrient. Glucose can only yield a large amount of energy when it is coupled to a powerful oxidizing agent, e.g., oxygen. It would be just as logical to consider oxygen as a "high energy" nutrient.

The ATP which is generated can be related to the quantity of cellular material synthesized. Bauchop and Elsden (1960) have shown that 10 g dry weight of cells are produced per mol ATP synthesized by substrate level phosphorylation.

1.1.1. Chemotrophic Oxidation Reduction Reactions

The significance of the term chemotrophic is seen in Fig. 1 where some other processes are also defined. In a more detailed consideration of oxidation/

reduction reactions (Table II) each reductant (electron donor) is considered against the various oxidants (electron acceptors). Only a few examples of each reaction are discussed in the following which may be considered a catalogue of the principal energy yielding chemotrophic reactions found in bacteria. All these reactions may be termed *dissimilative* since neither of the reactants are assimilated. The oxidation/reduction reactions are discussed in the order in which they are listed vertically in Table II, i.e., with increasing energy yield going downwards but decreasing energy yields in each series from left to right.

1. H_2 oxidations

Molecular hydrogen participation in biological reactions is mediated by hydrogenases. Hydrogen is highly reactive and rarely accumulates in any environment. It is produced in significant quantities only in anaerobic situations. It is largely used by methane producers but some bacteria may oxidize it by alternative mechanisms as shown by the following examples.

H_2/CHO   The following process has been described for *Vibrio succinogenes* (Iannotti *et al.*, 1973):

$$\underset{\text{fumarate}}{COOHCHCHCOO^-} + H_2 \rightleftharpoons \underset{\text{succinate}}{COOHCH_2CH_2COO^-}$$

$$\Delta G'_o = -20{\cdot}6 \text{ kcal.}$$

This type of reaction probably occurs infrequently, only when high concentrations of suitable CHO e^- acceptors are present with H_2, a combination which is not common.

H_2/CO_2 The process:

$$H_2 + \tfrac{1}{4}CO_2 \rightleftharpoons \tfrac{1}{4}CH_4 + \tfrac{1}{2}H_2O; \qquad \Delta G'_o = -8{\cdot}3 \text{ kcal,}$$

is carried out by all methanogenic bacteria (Wolfe, 1972) and is quantitatively the most important process involving H_2 in non-sulfate environments. In *Clostridium aceticum* the following process may take place:

$$H_2 + \tfrac{1}{2}CO_2 \rightleftharpoons \tfrac{1}{4}CH_3COO^- + \tfrac{1}{2}H_2O; \qquad \Delta G'_o = -4{\cdot}3 \text{ kcal,}$$

(Wieringa, 1940; Mah *et al.*, 1976). Although this reaction is less favorable than methane production, it may occur preferentially in sewage sludge. The acetate is then converted into methane by other bacteria (Mah *et al.*, 1976).

H_2/SO_4^{2-} The oxidation with sulfate is carried out by *Desulfovibrio* spp. (Postgate, 1969):

$$H_2 + \tfrac{1}{4}SO_4^{2-} + \tfrac{1}{4}H^+ \rightleftharpoons \tfrac{1}{4}HS^- + H_2O; \qquad \Delta G'_o = -9{\cdot}1 \text{ kcal.}$$

The oxidation by sulfate yields more energy than the oxidation by carbon dioxide and presumably occurs preferentially when both oxidants are present.

H_2/NO_3^- The process:

$$H_2 + 2/5NO_3^- + 2/5H^+ \leftrightharpoons 1/5N_2 + 6/5H_2O;$$

$$\Delta G_o' = -53 \cdot 6 \text{ kcal},$$

has been described for *Paracoccus* (*Micrococcus*) *denitrificans* and *Alcaligenes* (*Hydrogenomonas*) *eutrophus* (Schlegel, 1975). It yields considerable free energy but hydrogen and nitrate may never reach significant concentrations in the same environments.

H_2/O_2 The two above mentioned bacteria may also carry out the reaction:

$$H_2 + \tfrac{1}{2}O_2 \leftrightharpoons H_2O; \quad \Delta G_o' = -56 \cdot 7 \text{ kcal}.$$

The probability is slight that significant hydrogen oxidation occurs by this process in nature.

2. *CHO oxidations*

These oxidations are performed by *heterotrophic bacteria*. The oxidations are important since a major portion of the dissimilatory carbon cycle flows through them. The reactions are very diverse and many different types of bacteria perform them. The following series illustrates particularly well the increasing energy yields per substrate carbon as one progresses down the series (increases in $\Delta \varepsilon$).

CHO/H_2O The following process is carried out by *Desulfovibrio* according to Bryant (1969):

$$\underset{\text{lactate}}{CH_3CHOHCOO^-} + H_2O \leftrightharpoons \underset{\text{acetate}}{CH_3COO^-} + CO_2 + 2H_2;$$

$$\Delta G_o' = +5 \cdot 7 \text{ kcal}.$$

The energy yield is negative and the bacteria can only grow when hydrogen is removed by coupling it to methane production. This also applies to the reaction:

$$\underset{\text{ethanol}}{CH_3CH_2OH} + H_2O \leftrightharpoons \underset{\text{acetate}}{CH_3COO^-} + 2H_2;$$

$$\Delta G_o' = +8 \cdot 3 \text{ kcal},$$

carried out by the "S-organism" (Bryant *et al.*, 1967). In a mixed culture Bryant *et al.* (1967), observed the reaction:

$$CH_3CH_2COO^- + 2H_2O \rightleftharpoons CH_3COO^- + CO_2 + 3H_2;$$

propionate   acetate

$$\Delta G'_o = +25 \cdot 1 \text{ kcal.}$$

This is again an example of a process where an apparently unfavorable thermodynamic oxidation can occur when one of the products, in this case hydrogen, is removed from the system. Wolin (1974) illustrates this in the reaction:

$$NADH + H^+ \rightleftharpoons NAD^+ + \tfrac{1}{2}H_2.$$

When the hydrogen partial pressure is 1 atm the K is $6 \cdot 7 \times 10^{-4}$ and $\Delta G'_o$ is $+4 \cdot 33$ kcal; when the partial pressure of hydrogen is reduced to 10^{-6} atm, the equilibrium constant is $6 \cdot 7 \times 10^2$ and $\Delta G'_o$ is $-3 \cdot 83$ kcal (at the specified pressure). Thus, at very low hydrogen pressures the production of hydrogen from NADH is favored. Even when hydrogen production is not coupled to phosphorylations, this type of reaction where NADH is oxidized by H^+ may be very important in providing a mechanism for NADH oxidation, without the necessity of utilizing a portion of the substrate, as must occur in typical fermentations, in order that the supply of the electron acceptor NAD^+ be renewed.

CHO/CHO   This type of oxidation, where part of the substrate is oxidized at the expense of a part of the substrate being reduced, ATP being generated by substrate level phosphorylation, is termed *fermentation*. The extent to which substrate level phosphorylation is involved in all the following reactions is not clear. The most common type of fermentation is that of sugars. This is carried out by a variety of bacteria, among them species of *Bacteroides*, *Escherichia*, *Ruminococcus*, *Bacillus*, *Clostridium*, and *Lactobacillus*. They produce a variety of end-products: carbon dioxide, formate, acetate, propionate, butyrate, lactate, hydrogen and alcohols, with a range of ATP yields. In mixed cultures (e.g., rumen, sewage sludge) the products are the simple volatile acids plus methane and carbon dioxide. This is partly due to the other products being re-metabolized but it is also due to the fermentation products of individual bacteria being altered by the mechanisms suggested above. In this way, through hydrogen evolution, the necessity to use substrate as an electron sink in order to re-oxidize NADH is avoided. The more reduced CHO products, ethanol, lactate, succinate, butyrate, butanediol, and propionate are absent.

A fermentation of glucose by a typical fermenting anaerobe such as *Lactobacillus*, results in lactate production. This fermentation should yield only two ATP mol^{-1} and the growth yields are consistent with this value. Many other anaerobic fermenters have higher growth yields than is consistent with ATP generation by substrate phosphorylations, perhaps suggesting additional phosphorylation reactions. Some anaerobes can assimilate up to 30% of the substrate carbon into cells.

Among other types of fermentations, in addition to those of sugars, may be mentioned butyric acid fermentation carried out by *Clostridium kluyveri*;

$$CH_3COO^- + CH_3CH_2OH \rightleftharpoons CH_3CH_2CH_2COO^- + H_2O;$$

acetate ethanol butyrate

$$\Delta G'_o = -9.2 \text{ kcal}$$

(Barker, 1956; Thauer *et al.*, 1968). This is a very low energy yield from which to make ATP yet the bacterium can grow. This is a good example of how fermentation products acetate and ethanol can be further metabolized to yield energy. The fermentation of acetate carried out by *Methanosarcina barkeri* (Mah *et al.*, 1976) takes place according to:

$$CH_3COO^- \rightleftharpoons CH_4 + CO_2; \qquad \Delta G'_o = -6.6 \text{ kcal}.$$

This has been found to be a significant pathway for the methane production in sewage sludge. Presumably substrate level phosphorylation cannot account for ATP generation in this reaction. If fermentation is defined as being associated with substrate level phosphorylation, perhaps this should not be called fermentation.

Amino acid fermentation (Strickland reaction) is carried out by various species of *Clostridium* (Barker, 1956). An example is:

$$CH_3CHNHCOOH + 2CH_2NHCOOH \rightleftharpoons 3CH_3COO^-$$

alanine glycine acetate

$$+ 3NH_4^+ + CO_2; \qquad \Delta G'_o = -147.9 \text{ kcal}.$$

Alanine is the electron donor, being oxidized to acetyl CoA

via pyruvate. Acetyl CoA can yield ATP by substrate level phosphorylation. This type of reaction is probably only important where protein degradation is occurring in quantity.

CHO/CO$_2$ A reaction carried out by *Clostridium thermoaceticum* is according to Andreesen *et al.* (1973):

$$C_6H_{12}O_6 \rightleftharpoons 2CH_3COOH + 2CO_2 + 4H. + (4ATP),$$

$$4H. + 2CO_2 \rightleftharpoons CH_3COOH.$$

The cell yield from the calculated ATP generated, a total of three since one is used in the second reaction, is higher than one would have expected. It is suggested that electron transport phosphorylation may have generated additional ATP from some step in the carbon dioxide reduction, although thermodynamically this would not appear to be favorable. Even without additional phosphorylations, the process of carbon dioxide reduction as an electron sink, represents a saving of e.g., pyruvate being lost as an energy source to the cells. This is the same principle as the use of hydrogen gas formation as an electron sink.

CHO/SO$_4^{2-}$ *Desulfovibrio* and *Desulfotomaculum* spp. carry out this sort of process (Postgate, 1969):

$$CH_3CHOHCOO^- + \tfrac{1}{2}SO_4^{2-} + 3/2H^+ \rightleftharpoons CH_3COO^-$$

$$+ CO_2 + H_2O + \tfrac{1}{2}HS^-; \qquad \Delta G'_o = -8.9 \text{ kcal.}$$

This type of reaction is only important in sulfate-rich, anaerobic environments where lactate is available from other fermentations. *Desulfovibrio* species cannot oxidize acetate further. *Desulfotomaculum acetoxidans*, however, catalyzes the reaction:

$$CH_3COO^- + SO_4^{2-} \rightleftharpoons 2CO_2 + 2H_2O + HS^-;$$

$$\Delta G'_o = -9.7 \text{ kcal.}$$

These bacteria have recently been isolated (Widdel and Pfennig, 1977); they are probably very important, particularly in anaerobic, sulfate-rich marine sediments. They are also able to oxidize a limited range of other carbon compounds.

CHO/S°   *Desulfuromonas acetoxidans* can oxidize acetate with sulfur according to:

$$CH_3COO^- + 2H_2O + 4S° \rightleftharpoons 2CO_2 + 4HS^- + 3H^+;$$

$$\Delta G'_o = -6\cdot0 \text{ kcal.}$$

(Pfennig and Biebl, 1976). This is rather similar to the oxidation carried out by sulfate. *Desulfuromonas acetoxidans* can also oxidize ethanol and propanol. The acetate oxidation can alternatively utilize malate or fumarate as electron acceptor, thus also putting the species in the category of forms utilizing CHO/CHO couples.

CHO/NO_3^- Species of *Escherichia*, *Bacillus*, *Proteus*, and many other forms are capable of facultatively anaerobic growth using NO_3^- as a terminal electron acceptor according to:

$$C_6H_{12}O_6 + 24/5NO_3^- + 24/5H^+ \rightleftharpoons 6CO_2 + 12/5N_2$$

$$+ 42/5H_2O; \qquad \Delta G'_o = -649 \text{ kcal.}$$

Not all these bacteria perform the complete reduction to dinitrogen, the intermediate product is often nitrite. This dissimilative nitrate reduction occurs only in the absence of oxygen, which is the preferred electron acceptor. Nitrate can replace oxygen in many types of oxidations with the exception of reactions involving oxygenases. The role of nitrate as terminal electron acceptor (respiratory dissimilation) is different from its function in some clostridial reactions where ammonia is the end product. These are known as *nitrate fermentations*. These involve oxidations of reduced NAD (see Section 5.3.1).

CHO/O_2 These processes are carried out by the aerobic respiring bacteria which metabolize, e.g., sugars according to:

$$C_6H_{12}O_6 + 6O_2 \rightleftharpoons 6CO_2 + 6H_2O; \qquad \Delta G'_o = -686 \text{ kcal.}$$

The aerobes comprise a very important group of bacteria which due to the favorable yield of energy, predominate in oxic environments. In addition, some forms can use molecular oxygen to initiate the degradation of aromatic compounds and hydrocarbons (Gibson, 1968).

Possibly some aerobic bacteria can grow on carbon monoxide as an energy source but these bacteria have not yet been investigated (Quayle, 1972). Some methanogenic bacteria can grow on CO (Daniels *et al.*, 1977):

$$4CO + 2H_2O \rightleftharpoons 3CO_2 + CH_4;$$

$$\Delta G'_o = -50 \cdot 5 \text{ kcal mol}^- \quad CH_4$$

3. CH_4 oxidations

Methane is oxidized principally by molecular oxygen, an oxygenase being necessary for the first attack on the methane molecule. This may explain the fact that no nitrate-utilizing bacterium can oxidize methane.

CH_4/SO_4^{2-}　Oxidation of methane by sulfate would proceed according to:

$$CH_4 + SO_4^{2-} \rightleftharpoons CO_2 + 2H_2O + HS^-; \qquad \Delta G'_o = -3 \cdot 1 \text{ kcal}.$$

There is some doubt that bacteria can grow on methane alone but there is evidence that methane oxidation does take place in anoxic, sulfate-rich sediments and basins.

CH_4/O_2　Species of *Pseudomonas*, *Methylomonas*, *Methylobacter*, and *Methylococcus* oxidize methane according to:

$$CH_4 + 2O_2 \rightleftharpoons CO_2 + 2H_2O; \qquad \Delta G'_o = -193 \cdot 5 \text{ kcal},$$

(Ribbons *et al.*, 1970).

4. HS^- oxidations

Not only sulfide, but also thiosulfate ($S_2O_3^{2-}$) and elemental sulfur may be oxidized by bacteria in order to generate energy. In all cases only nitrate or oxygen are sufficiently powerful oxidants to generate biologically useful energy.

HS^-/NO_3^-　*Thiobacillus denitrificans* utilizes the reaction:

$$HS^- + 8/5NO_3^- + 3/5H^+ \rightleftharpoons SO_4^{2-} + 4/5N_2 + 4/5H_2O;$$

$$\Delta G'_o = -177 \cdot 9 \text{ kcal}.$$

The bacterium can also use oxygen as an electron acceptor.

HS^-/O_2　The "white sulfur bacteria", e.g., species of *Thiobacillus*, *Beggiatoa*, *Thioploca*, *Achromatium*, and *Thiovolum* carry out the process (Schlegel, 1975):

$$HS^- + 2O_2 \rightleftharpoons SO_4^{2-} + H^+; \qquad \Delta G'_o = -190 \cdot 4 \text{ kcal}.$$

The biochemistry of the process is complex and involves an

unusual compound APS (adenosine 5'-phosphosulfonate). It is also the only lithotrophic process which involves substrate level phosphorylation.

5. Oxidations of inorganic N-compounds

Only oxygen can serve as an electron acceptor for oxidizing N-compounds and at the same time yield biologically useful energy.

NH_4^+/O_2 Ammonia oxidation is carried out by species of *Nitrosomonas*, *Nitrosocystis*, *Nitrosospira*, and *Nitrosolobus* (Schlegel, 1975) according to:

$$NH_4^+ + 3/2O_2 \rightleftharpoons NO_2^- + 2H^+ + H_2O;$$

$$\Delta G_o' = -65.7 \text{ kcal.}$$

This is the first step in *nitrification*. The nitrite is further oxidized, as shown below.

NO_2^-/O_2 Species of *Nitrobacter*, *Nitrococcus*, and *Nitrospina* (Schlegel, 1975) complete the nitrification by oxidizing nitrite according to:

$$NO_2^- + \tfrac{1}{2}O_2 \rightleftharpoons NO_3^-; \qquad \Delta G_o' = -18.1 \text{ kcal.}$$

Dinitrogen oxidation is not carried out by any known organism. Broda (1975b) points out that the reaction:

$$N_2 + 5/2O_2 + H_2O \rightleftharpoons 2NO_3^- + 2H^+; \qquad \Delta G_o' = -15.6 \text{ kcal.}$$

is thermodynamically favorable. Perhaps no bacteria are capable of utilizing the energy available from it because of the stability of dinitrogen. The high activation energy involved in dissociating the N_2 molecule is also reflected in the difficulty in performing nitrogen fixation.

6. Oxidation of ferrous iron

Fe^{2+}/O_2 *Thiobacillus ferrooxidans* can grow on the basis of the energy yield of the process:

$$Fe^{2+} + \tfrac{1}{4}O_2 + H^+ \rightleftharpoons Fe^{3+} + \tfrac{1}{2}H_2O;$$

$$\Delta G_o' = -1.04 \qquad \Delta G° = -10.6 \text{ kcal.}$$

($\Delta G_o'$ for the process is only -1.04 kcal but since the bacteriologically mediated process only takes place at very low pH (see Section 7.1.1) this value has no special meaning.) Several other bacteria seem to catalyze the oxidation of reduced iron (and manganese) but this does not seem to have bioenergetic significance.

1.1.2. Phototrophic Reactions

Instead of chemotrophic reactions, some organisms (plants and photo-synthetic prokaryotes) can utilize the energy of light. With respect to prokaryotes it is convenient to distinguish between oxygenic and anoxygenic reactions.

Oxygen producing reactions. All phototrophic phosphorylations are mediated through electron transport and presumably the generation of a proton gradient. The essentials of the process in green plants and blue-green bacteria (which represent oxygenic photosynthesis) are illustrated in Fig. 3. Chlorophyll is activated by photons and an electron is energized to a higher state, this electron is accepted by a low potential acceptor, from which it flows down an electron transport chain, analogous to the respiratory chain illustrated in Fig. 2. It loses energy, as it goes from carrier to carrier and finally reaches chlorophyll at its original energy level. Some of the energy that it loses is conserved as ATP. Some ATP is also generated by noncyclic processes.

Non-oxygen evolving reactions. Anaerobic bacterial photophosphorylations are exclusively cyclic but resemble in other respects the blue-green system (Fig. 3). The anaerobic bacterial systems are chiefly different in that a whole range of electron donors are used to donate reducing power for carbon dioxide reduction and assimilation. These electron donors are dissimilatively oxidized and the products of H_2, CHO (acetate, etc.), CH_4, HS^-, S°, are the same as in the chemotrophic systems which have been discussed. No bacterium appears to use ammonia as an electron donor.

Rhodopsin photophosphorylation. Bacterio-rhodopsin is a carotene pigment present in the membrane of the halophile (salt-loving) bacterium *Halobacterium halobium* (Oesterhelt *et al.*, 1977). This purple pigment is activated by light with the displacement, across the cytoplasmatic membrane, of protons. This generates a potential across the membrane due to the higher proton concentration on the outside. These protons are driven through membrane-located ATPase, synthesizing ATP, as illustrated in Fig. 4. The process operates when conditions are unfavorable for energy generation through oxygen respiration. Bacterio-rhodopsin does not seem to occur widely in bacteria and may not be quantitatively significant in energy generation, except for halophiles.

1.2. PRECURSORS REQUIRED FOR CELL SYNTHESIS

Bacteria require a variety of materials for the synthesis of all cellular components so that they may grow and finally divide to produce progeny.

Growth may be defined by reference to this balanced increase in biomass, and if bacteria can be said to have a purpose, it is to increase in size and divide. Most bacterial activities are directed to this end, and one of the most important related processes is that of acquiring the necessary components from the environment for the synthesis of cellular material. The gross requirements are dictated by the elemental composition of the cell (Table III). A major portion of the cell is composed of C (55%) and N (10%), and these together with oxygen, hydrogen, phosphorus, and sulfur are the constituents of the amino acids, nucleotides, sugars, and fatty acids which, when polymerized, yield the macromolecules: protein, nucleic acid, polysaccharide, and lipid, which together constitute 97.3% of the cell dry weight (Table IV).

In an environment which contains free amino acids and sugars, a bacterium can readily assimilate these building blocks for macromolecular synthesis. Such environments are rare and are restricted to live macroorganisms or

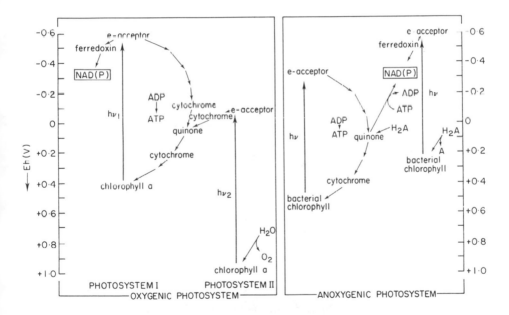

Fig. 3. Comparison of cyanobacterial and plant, oxygenic (left) and bacterial, anoxygenic (right) photosynthesis. The illustrations are simplified to show the main processes which involve the production of NAD(P)H and ATP, to be used for CO_2 reduction. In the oxygenic photosynthesis, ATP is generated by cyclic photophosphorylation in photosystem I and NAD(P)H is also produced by this photosystem. The electrons thus diverted to $NAD(P)^+$ are replaced from photosystem II. In the anoxygenic system ATP is also generated by cyclic photophosphorylations but NAD(P)H cannot be made directly from this first photosystem. Electrons must first be activated by either expenditure of ATP or by another photosystem.

their decomposition products. There is intense competition for these niches and the majority of bacteria are forced to assimilate one or more of the essential elements from non-organic, mineral sources. The main mineral forms in which C, N, S, and P occur are shown in Table V. These elements most often occur in an oxidized form in the environment, they must, therefore be reduced to the proper state before they can be utilized for synthesis. This

TABLE III. Elemental Composition of Bacteria.

Element	% of dry weight
C	55
O	20
N	10
H	8
P	3
S	1

TABLE IV. Composition of Bacteria.

Monomer	Elements	Polymer	% Polymer[a]
Amino acid	CHNOS	Protein	52·4
Organic bases	CNOHP	Nucleic acid	19·9
Sugar	CHO	Polysaccharide	16·6
C-16 acid + P	CHOP	Phospholipid	9·4
		Total	97·3

[a] Data (as % of dry weight) from Stouthamer (1973).

TABLE V. Principal Mineral Forms of C, N, S, and P.

Element	In oxic environments	In anoxic environments
C	CO_2	CO_2, CH_4
N	NO_3^-, NO_2^-, N_2	N_2, NH_4^+
S	SO_4^{2-}	$S^{\cdot \cdot}$, SH^-
P	PO_4^{3-}	PO_4^{3-}

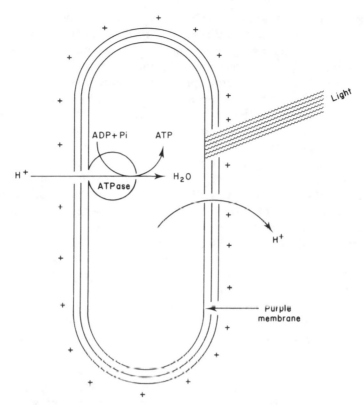

Fig. 4. Light energy conversion in Halobacteria. Bacterio-rhodopsin, a purple membrane of *Halobacterium* species, absorbs light and a proton is liberated to the medium. In the light-driven regeneration of the purple complex, a proton is taken up from the cytoplasm to be passed through the cytoplasmic membrane. The proton gradient thus created drives an ATPase in reverse, generating ATP from ADP and inorganic phosphate.

reductive assimilation will be discussed for each of the elements to illustrate the diverse types of bacteria capable of utilizing the diverse minerals that are available to them. It should be noted that although these reactions, which involve reductive assimilations (the addition of one or more electrons) are similar to energy yielding oxidation/reduction reactions, they are not coupled to ATP generation. Electrons which could have been used in energy-yielding reactions are thus lost and constitute an energy drain for the organism.

1.2.1. Carbon Assimilation

Microorganisms can be divided into two groups based on their carbon assimilatory characteristics; the *heterotrophs* and the *autotrophs* (Fig. 1). The assimilation of reduced organic carbon compounds (heterotrophic growth) is the most usual method for microorganisms to obtain the material for cell synthesis, and a discussion of the diverse capabilities of microorganisms in assimilating organic compounds is not necessary. It is sufficient to say that a naturally occurring, low molecular weight, organic compound will be taken up by heterotrophic cells. It should be noted that the concept of heterotrophy implies the use of organic carbon compounds for both energy yielding reactions and for assimilation. Carbon dioxide is the mineral form of carbon that is available in virtually every environment, in non-limiting quantities. Its reductive assimilation is, however, restricted to plants and a limited group of bacteria, known as autotrophs. An autotroph may be defined as an organism with the capacity to use carbon dioxide, by reductive assimilation, for most of its carbon requirements. Some microorganisms are facultatively autotrophic, they reduce CO_2 when no organic carbon is available but can grow heterotrophically when it is available. There has been considerable interest in the underlying biochemical basis for obligate autotrophy, but no single factor appears to be capable of explaining it and it has no particular relevance to the present discussion. Reduction of carbon dioxide is an energy expensive process, since as we have discussed, electrons have to be used which would otherwise pass down an electron transport chain with coupled oxidative phosphorylations. This process of CO_2 reduction cannot be coupled to ATP generation as the synthetic process must be driven by a large decrease in the free energy. Such a process is best performed by photosynthetic organisms, to whom radiant energy is readily available. It is, however, also a property of some chemotrophic bacteria. Heterotrophic bacteria (oxidation of organic carbon) are not autotrophic as this would be a process involving the oxidation of one organic compound so the carbon dioxide could be assimilated. Many lithotrophic bacteria (oxidation of inorganic compounds) are autotrophic.

It is characteristic of many autotrophic microorganisms to possess complex invaginations of the cytoplasmic membrane. These are found not only in many photosynthetic bacteria, ammonia-, and nitrite-oxidizing bacteria, methane-producing bacteria, but also in the "autotrophic" methane oxidizing bacteria. It is likely that this is a modification to create more reaction sites for energy coupling. Heterotrophic bacteria, both fermenters and respirers, also assimilate carbon dioxide through a variety of biochemical processes for incorporation into cells or products. The process, while it is often essential, does not supply more than about 6% of the cell carbon. Kuznetzov (1968) has used this process as an index of heterotrophic cell synthesis (see also Section 3.1.3).

Photoautotrophs. Green plants represent the best known type of photoautotrophy. Figure 3 illustrates the most important features of this process,

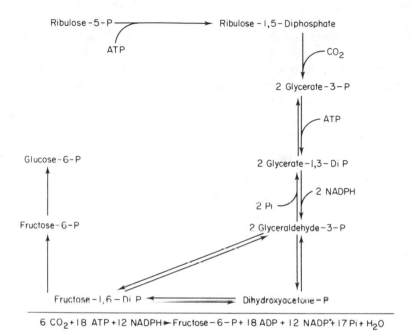

$$6\ CO_2 + 18\ ATP + 12\ NADPH \rightarrow Fructose-6-P + 18\ ADP + 12\ NADP^+ + 17\ Pi + H_2O$$

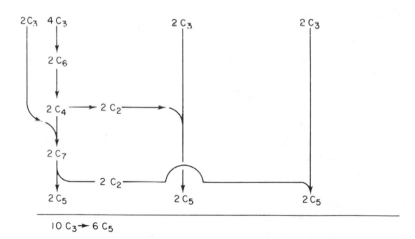

$$10\ C_3 \rightarrow 6\ C_5$$

Fig. 5. The reductive ribulose diphosphate pathway showing the reduction of CO_2 (top) and the process by which ribulose diphosphate is regenerated (bottom). Ten out of every twelve triose molecules (C_3) generated from CO_2 reduction are used in the production of six ribulose molecules (C_5).

namely the generation of ATP by cyclic and non-cyclic phosphorylation, by light-activated electrons, the generation of reduced NADP by other light-activated electrons and the utilization in a non-light reaction of ATP and reduced NADP (Fig. 5) to reduce CO_2. In addition to ATP and reduced NADP, oxygen is an end-product of the light reaction. The blue-green bacteria are the only prokaryotes capable of this type of oxygen-generation photosynthesis. There are, however, other prokaryotes capable of photo-autotrophy, but using compounds other than water as the electron donors. Both types of photosynthesis may be summarized as follows:

$$CO_2 + A \text{ red (H donor)} \rightarrow (CH_2O) + A \text{ ox} + H_2O.$$

A would represent H_2O in plant-type photosynthesis, producing O_2; it would represent H_2S or H_2 in anaerobic-type photosynthesis producing elemental $S°$ or H_2O. The oxidation of elemental $S°$ can proceed further with the production SO_4^{2-}. A characteristic of non-oxygen producing photo-autotrophs is that they are anaerobes. These forms belong to three groups: the purple sulfur bacteria (Chromatiaceae), the green sulfur bacteria (Chlorobiaceae), and the purple non-sulfur bacteria (Rhodospirillaceae). The main function of the photoprocess is the generation of ATP. Low potential (high energy) electrons for reducing CO_2 are probably produced by an ATP-driven reverse electron flow or by a non-cyclic process in at least some bacteria, e.g. *Chromatium* (Morita, 1968) and *Rhodospirillum rubrum* (Sybesma, 1969). Most of the Chlorobiaceae are obligate autotrophs (absolute CO_2 requirement). The Rhodospirillaceae and some Chromatiaceae are facultative autotrophs able to assimilate carbon compounds. Some of the Rhodospirillaceae, e.g. *Rhodospirillum*, can also grow aerobically, hetero-trophically, in the dark.

All these different photoautotrophs have in common the Calvin reductive ribulose diphosphate pathway (Fig. 5). Other less significant pathways may also be used, but these microorganisms always possess the key enzymes ribulose-1, 5-diphosphate carboxylase and ribulose-5-phosphokinase, when growing autotrophically. This is also true of the next group of autotrophs, the chemoautotrophs.

Chemoautotrophs. These microorganisms typically reduce CO_2 by the reductive Calvin cycle using ATP and reduced NADP both produced by the oxidation of reduced inorganic substrates. These are the chemo-lithotrophic autotrophs, their substrates being reduced inorganic nitrogen, sulfur and iron compounds or hydrogen. Their physiology has been reviewed recently by Schlegel (1975) an excellent source of further references. A typical lithotrophic autotroph possesses the following characteristics:

   (1)   It is an obligate autotroph, using the Calvin cycle.

(2) It generates ATP by electron transport phosphorylations with O_2 as terminal electron acceptor.

(3) It is unable to directly reduce $NADP^+$, and since NADPH is required for CO_2 reduction, this must be made by "reverse electron flow" driven by ATP generated by the normal flow of electrons to oxygen (Fig. 6).

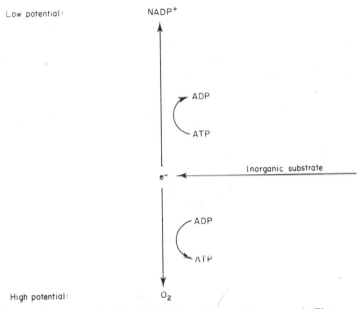

Fig. 6. Generation of NAD(P)H in a typical chemolithotrophic autotroph. Electrons from an inorganic substrate (e.g. H_2S, NH_3) cannot reduce $NADP^+$. In order to achieve this reduction ATP must be expended to "activate" the electrons.

Exceptions to this general description will be given in the following survey. The nitrifying bacteria are typical chemolithotrophic autotrophs. Those bacteria (*Nitrosomonas, Nitrosocystis, Nitrosospira*, and *Nitrosolobus*) which oxidize ammonia are all obligate autotrophs. This is also true of the nitrite oxidizing bacteria (*Nitrobacter, Nitrococcus*, and *Nitrospina*), except for some species of *Nitrobacter* which are not obligate autotrophs. The sulfur oxidizing autotrophic bacteria are a mixed group and autotrophy has not been well established for the filamentous genera *Beggiatoa, Thioploca*, and *Thiothrix*, or for the non-filamentous genera *Achromatium, Thiophysa*, and *Thiovolum*. Species of *Thiobacillus* are autotrophic, but many are not obligately so. They differ also from the general description in that some ATP may be generated by substrate level phosphorylation, and one species, *Thio-*

bacillus denitrificans, can utilize nitrate as an alternative electron acceptor instead of oxygen. All, however, conform to the pattern of "reversed electron flow" in reducing $NADP^+$.

The autotrophy of bacteria oxidizing ferrous iron has been proved only for *Thiobacillus ferrooxidans*. This bacterium may also oxidize sulfur or thiosulfate and conforms to the general description of a chemolithotrophic autotroph. The ability of other iron bacteria (*Gallionella, Leptothrix, Clado-thrix*, and *Crenothrix*) to reduce CO_2 is in doubt (see Section 7.1.2.). The hy-drogen oxidizing autotrophs are all facultative autotrophs, a property associated with their capacity to directly reduce $NADP^+$ by hydrogen oxida-tion (Smith *et al.*, 1967). Species of genus *Pseudomonas, Alcaligenes* (*Hydro-genomonas*), *Paracoccus, Nocardia, Brevibacterium*, and *Arthrobacter* oxidize hydrogen with oxygen; *Alcaligenes eutrophus* and *Paracoccus denitrificans* can use nitrate as an alternative electron acceptor, producing dinitrogen. There are some other bacteria which can oxidize hydrogen (*Desulfovibrio* and *De-sulfotomaculum*), but they cannot grow autotrophically even though they may be able to reduce some CO_2. There are yet more bacteria which for a variety of reasons, do not conform to the description of a typical lithotrophic autotroph. The following classification of these diverse metabolic types is based on Whittenbury and Kelly (1977). They are distinguished by not having a ribulose diphosphate (Calvin) cycle for CO_2 reduction, and many are technically not lithotrophic since they oxidize organic carbon compounds, typically the one carbon (C_1) compounds: methane (CH_4), methanol (CH_3OH), formaldehyde (HCHO), formic acid (HCOOH), or methyl-amine (CH_3NH_2).

Anaerobic carbon dioxide reducers. All methanogenic bacteria can grow litho-trophically on hydrogen and carbon dioxide, with the production of methane. They are thus chemolithotrophic autotrophs, but they do not use the ribulose-diphosphate cycle for carbon dioxide assimilation. Neither do they use an alternative ribulose monophosphate cycle, but they may use the "serine" pathway (Fig. 7). The carbon dioxide is reduced through various intermediates on the pathway to methane and probably one of these, for-maldehyde, is the carbon source actually assimilated for cell synthesis. Some species of the genus *Clostridium* can also grow lithotrophically on hydrogen and carbon dioxide producing acetic acid (Wieringa, 1940). The pathways of carbon dioxide reduction are unknown, but *Clostridium thermoaceticum*, growing heterotrophically, can reduce carbon dioxide to acetic acid via formate, using NADP produced during glycolysis.

Anaerobic methane oxidizers. This group of microorganisms oxidizes methane with sulfate, producing sulfide and carbon dioxide (Hanson, quoted by Whittenbury and Kelly, 1977). Different strains may have different pathways of assimilation, but one strain, in common with the methane producers, appears to incorporate formaldehyde.

Aerobic oxidizers which assimilate carbon dioxide. Paracoccus (Micrococcus) denitrificans can be a chemolithotrophic autotroph growing on hydrogen and carbon dioxide; it can also grow on methanol as an energy source. On this substrate it has an obligate requirement for carbon dioxide which is reduced by the ribulose diphosphate pathway and assimilated.

The same pattern of metabolism is seen for *Pseudomonas oxalaticus* growing on formate.

Aerobic C_1 oxidizers which assimilate C_1 compounds. The best investigated microorganisms in this group are the methane oxidizers, but it also includes yeasts, filamentous fungi, and species of *Bacillus, Caulobacter*, and *Assticacaulis*. None of these organisms reduce carbon dioxide as a major contributor to cellular carbon, and all oxidize the C_1-substrate to formaldehyde, which is assimilated through either the "Serine" or ribulose monophosphate pathways (Fig. 7). The obligate methane-oxidizers fall into two groups which use one or other of the assimilative pathways, but *Methylococcus capsulatus* possesses enzymes for both pathways. All methane-oxidizers require carbon dioxide as an additional essential carbon source.

1.2.2. Nitrogen Assimilation

Nitrogen constitutes a major portion of cells (Table III) occurring principally in proteins and nucleic acids (Table IV). The monomer units of these macromolecules (amino acids, purine and pyrimidine bases) are taken up and assimilated by most bacteria. Highly charged nucleotides cross the cytoplasmic membrane with more difficulty. Specific permeases are usually involved, and should a cell lack a particular permease it would encounter difficulties in obtaining sufficient amounts of nutrient from a dilute medium. Some bacteria prefer to take up oligopeptides in preference to amino acids (Pitman *et. al.*, 1967), while other bacteria use ammonia in preference to amino acids (Hullah and Blackburn, 1970). This preference for ammonia utilization is particularly true in anaerobic environments like the rumen, where extensive degradation of the carbon skeleton of amino acids occurs with the liberation of ammonia (Blackburn, 1965).

Ammonia assimilation. Ammonia is the predominant form in which inorganic nitrogen occurs in anaerobic environments. It is also found in aerobic environments, e.g. in soil, but it is in lower concentration than nitrate, the more oxidized form of nitrogen. Most bacteria are capable of synthesizing some or all of their amino acids and organic bases from ammonia.

The valency state of nitrogen in ammonia is the same as in organic nitrogen compounds, so no costly reductions are necessary. It is not necessary to discuss in detail the biosynthetic pathways involved; a short summary is given by Levy *et al.* (1973).

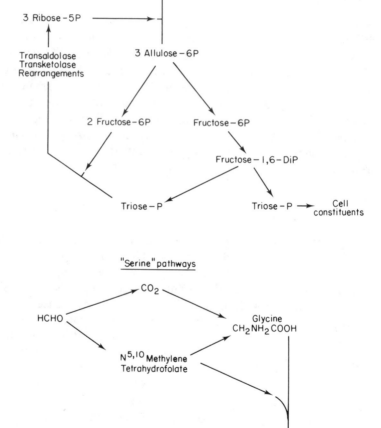

Fig. 7. The reductive monophosphate pathway (top) and the reductive serine pathway (bottom) showing the reduction of formaldehyde.

Dinitrogen assimilation (nitrogen fixation). In environments where organic nitrogen, ammonia, and nitrate are limiting, it is a considerable advantage to be able to assimilate dinitrogen (N_2) which is present in such large quantities in the atmosphere. The capacity to assimilate and reduce it to ammonia for incorporation into cell constituents is restricted to prokaryotes. Many unrelated groups prokaryotes possess this capacity (Table VI), but it is not possessed by a large total number of these microorganisms. Nitrogen fixation is dependent on the possession of the enzyme nitrogenase, which is made of two proteins. "Protein 1" contains molybdenum, non-heme iron and labile sulfur, has a molecular weight of approximately 2×10^5 Daltons, is not particularly sensitive to oxygen, and is known as molybdoferredoxin. "Protein 2" contains no Mo, non-heme iron, less labile sulfur, has a molecular weight of approximately 0.4×10^5 Daltons, is very sensitive to oxygen, and is known as azoferredoxin (Fig. 8). Nitrogenase is not highly specific for dinitrogen and is capable of reducing acetylene, cyanide, azide, nitrous oxide, and carbon monoxide. The reduction of acetylene to ethylene is the basis of the widely used acetylene-reduction assay for nitrogenase activity. Ethylene, the product, can be measured with great precision by gas chromatographic analysis. *In vitro* nitrogenase activity is dependent on Mg^{2+}, a reductant, and ATP.

TABLE VI. Representative Types of N_2-Fixing Bacteria.

	Phototrophic	Chemotrophic
Freeliving, oxic	Cyanobacteria	*Azotobacter* *Mycobacterium* Methane oxidizers *Thiobacillus*
Freeliving, anoxic	*Chromatium* *Chlorobium* *Rhodospirillum*	*Clostridium* *Klebsiella* *Bacillus* *Desulfovibrio* *Desulfotomaculum* Methanogenic bacteria
Symbiotic, oxic	Cyanobacteria (+ fungi, ferns *Cycas*, etc.)	*Rhizobium* (+ legumes, grass) *Spirillum* (+ grass) Unknown (+ alder, hawthorn, etc.)
Symbiotic, anoxic	Unknown	*Citrobacter* (+ termites)

This dependence of dinitrogen reduction on a reducing environment and on a plentiful supply of ATP is reflected in the type of microorganisms carrying out this reaction, as illustrated in Table VI (Brill, 1975; Stewart, 1976).

Freeliving, aerobic phototrophic bacteria. This group contains the blue-green bacteria. Because they are phototrophic they can generate unlimited ATP and reducing power, provided that sunlight is adequate. They do have a problem in maintaining an anoxic environment in which the oxygen-sensitive nitrogenase may act. Many blue-green bacteria have modified cells known as heterocysts, which contain nitrogenase and fix nitrogen. These cells have no photosystem 2 and are therefore incapable of generating oxygen, but the intact photosystem 1 produces ATP. Some blue-greens do not have the heterocysts and yet still fix nitrogen. In some cases at least this is due to the cells packing closely together to prevent the diffusion of oxygen and excess light reaching the semi-anaerobic nitrogen-reducing cells at the center (Kenyon *et al.*, 1972). Nitrogenase is present in many blue-green bacteria and they are consequently independent of N- and C-compounds. Cyanobacteria are often the first colonizers of nutrient-poor environments.

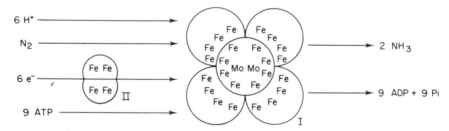

Fig. 8. Dinitrogen reduction by nitrogenase. Component II (two subunits with 2Fe each) accepts electrons via a transport chain from a donor (photosystem, reduced carbon, hydrogen etc.). These electrons are transferred to the complex component I (four subunits with 6Fe each, two of which are shared with the molybdenum cofactor) and then transferred to dinitrogen. ATP is expended in the activating of the reaction in an unknown manner. Protons are picked up by the reduced nitrogen to form ammonia which is incorporated into amino acids.

Freeliving, anaerobic, phototrophic bacteria. This group of bacteria has all the factors necessary for nitrogen fixation: a light driven ATP generating system, a reduced environment, and no problem with oxygen generation and toxicity. They have not been systematically studied, but nitrogenase is found in members of the three groups, the Chromatiaceae, the Rhodospirillaceae, and the Chlorobiaceae. The distribution of these anaerobic, phototrophic, nitrogen-fixers is very limited, and usually the environment contains adequate ammonia; it is therefore difficult to assess the ecological significance of their nitrogen-fixing ability. It seems likely that nitrogenase is widespread in the group.

Freeliving, aerobic chemotrophic bacteria. This group contains a variety of bacteria which are dependent on oxygen, but whose nitrogen-fixing ability is easily inhibited by too high an oxygen tension. *Azotobacter* species protect the nitrogenase by adopting a very high rate of respiration to regulate oxygen tension. This is very wasteful on substrate, and usually the bacteria are energy limited. *Azotobacter* has another protective system which allows the nitrogenase to undergo a conformational change to an inactive but oxygen-resistant form. In general, aerobic nitrogen-fixing bacteria possess mucoid capsules, which appear to function as barriers to excess oxygen reaching the cell. Mucoid material is also found in the blue-green bacteria and some of the symbiotic forms. The aerobic respirers are at a disadvantage in being more energy (ATP) limited than the phototrophs; the chemolithotrophs are at an even greater disadvantage since they gain little energy from their oxidations. One strain of *Thiobacillus ferrooxidans*, however, fixes nitrogen (Mackintosh, 1971).

Freeliving, anaerobic, chemotrophic bacteria. This group, like the anaerobic phototrophs, has no problem in maintaining a reduced environment for the nitrogenase but, unlike them, suffers from an energy limitation except under exceptional circumstances. A diversity of substrates are used: sugars (*Clostridium, Klebsiella, Bacillus*), hydrogen, and lactate (*Desulfovibrio, Desulfotomaculum*).

Symbiotic, aerobic, phototrophic bacteria. Lichens are interesting associations between blue-green bacteria and fungi. Presumably the blue-green bacteria fix nitrogen when the lichen is located on a rock, deficient in nitrogen. The fungus would benefit from the products of nitrogen and carbon dioxide reduction. It is more difficult to see how the bacterium benefits, except by being provided with an anchor to the rock, mineral solubilization, and possibly a regulation of oxygen tension. Blue-green bacteria characteristically are found associated with other heterotrophic bacteria, presumably the association is beneficial to both. Blue-green bacteria also form associations with plants: liverworts, mosses, ferns, gymnosperms and angiosperms (Section 8.2.3).

Symbiotic, aerobic, chemotrophic bacteria. This group of bacteria has solved the problem of energy supply by tapping the reduced carbon supply of plants. These associations range from the loose adhesion of *Rhizozobium paspalum* to tropical grass roots (Döbereiner, 1974) and *Spirillum lipoferum* to maize roots (Day *et al.*, 1975), to very specific nodule formation by *Rhizobium* species with leguminous plants. New nodule-forming associations are discovered each year (Nutman, 1976), but there has been little success in culturing these species and examining the nitrogen-fixing capabilities. *Rhizobium* species are readily cultured, and it has recently been shown that they can fix nitrogen in pure culture, provided that there is a controlled, low oxygen tension (McComb, 1975, see also Sections 8.2.1 and 8.2.2).

Symbiotic, anaerobic, phototrophic bacteria. There are not well documented cases of this type of association.

Symbiotic, anaerobic, chemotrophic bacteria. Examples of this are not well known, but some nodule-bacteria and forms associated with termites may be of this type (Section 8.2.4).

Nitrate assimilation. Nitrate is the predominant form in which inorganic N occurs in oxic environments. It is probably a minor source for nitrogen assimilation by anaerobic microorganisms. All plants and many aerobic bacteria possess the capability of nitrate assimilation. Nitrate is reduced through the following process:

$$\underset{+5}{\overset{NO_3^-}{nitrate}} \longrightarrow \underset{+3}{\overset{NO_2^-}{nitrite}} \longrightarrow \underset{+1}{?} \longrightarrow \underset{-1}{\overset{NH_2OH}{hydroxylamine}} \longrightarrow \underset{-3}{\overset{NH_4^+}{ammonia}}$$

Nitrate reductase is a molybdoflavoprotein which accepts electrons from NADPH and reduces nitrate to nitrite. The $+1$ intermediate in the reductive pathway to ammonia has not been identified. The further reductions are coupled to the oxidation of NADH or NADPH with no coupled phosphorylations. The reductions are thus costly in potential energy-yielding NAD(P)H.

1.2.3. Sulfur Assimilation

Sulfur is seldom a limiting nutrient for bacteria and is usually present as hydrogen sulfide or as sulfate in anaerobic environments and as sulfate in aerobic ones. It is incorporated into proteins as sulfide. Plants and aerobic bacteria must therefore reduce it as follows:

$$\underset{+6}{\overset{SO_4^{2-}}{sulfate}} \longrightarrow \underset{+4}{\overset{SO_3^{2-}}{sulfite}} \longrightarrow \underset{-2}{\overset{SH^-}{sulfide}}$$

The sulfate is activated by ATP, forming APS (adenosine-5' phosphosulfonate)

$$ATP + SO_4^{2-} \longrightarrow APS + PPi.$$

The pyrophosphate (PPi) is hydrolyzed by pyrophosphatase driving the reaction. This is similar to the activation of sulfate prior to dissimilative reduction (Section 1.1.1). The next activation step to PAPS (3'-phosphoadenosine-5'-phosphosulfonate) is unique to the assimilative reaction. The reduction of PAPS proceeds to sulfite. There is considerable expenditure of ATP and NADPH in the activation of sulfate and its reduction to sulfite. More NADPH is used up in the reduction of sulfite to sulfide:

$$SO_3^{2-} + 3NADPH + 4H^+ \longrightarrow HS^- + 3NADP^+ + 3H_2O.$$

The complete eight electron reduction is thus expensive in low potential reducing power and ATP. None of the NADPH reductions are coupled to ATP generation.

1.2.4 Phosphorus Assimilation

All life forms have a requirement for phosphorus and this demand is usually met by the assimilation of inorganic phosphate. It is incorporated into cells without any valency change (see also Section 7.2).

1.2.5. Other Assimilations

Bacteria with cytochromes and ferredoxin-type proteins have a requirement for iron which may be difficult to meet since the iron will often be found in the form of insoluble ferric hydroxide or ferrous sulfide complexes. The oxidation or reduction of iron is easily accomplished and constitutes little change in free energy.

Magnesium requirements are met by adequate supplies of Mg^{2+}, no reduction is necessary. There are minor requirements for Mo, Co, Zn, Cu, Ca, and Mn.

CHAPTER 2

The Evolution of Prokaryotic Metabolic Pathways

A consideration of the early evolution of life, which is really prokaryotic evolution, is relevant in the context of this book since biological evolution has had a very important impact on the present composition of the atmosphere, lithosphere and biosphere. The parallel changes in the spectrum of prokaryotic types and in the chemical environment of the earth are discussed in two sections, 2.1 Biological Evidence and 2.2 Geological Evidence.

2.1. BIOLOGICAL EVIDENCE

The scheme for the evolution of metabolic pathways has been based principally on the pathways that are now known to exist in bacteria, and by evaluating these in degrees of primitiveness. There are other lines of evidence which are considered where applicable. Similarities in the amino acid sequences in proteins is a useful method for establishing relatedness between polypeptides and thus between the bacteria which produce them. At a higher level of resolution the relatedness of DNA molecules can be compared. This has not been applied on a significant scale to the problem of prokaryotic evolution. A low percentage of guanosine and cytosine is indicative of primitiveness, but this may not be universally true.

The evolution of bacteria and hence the evolution of their metabolic pathways has been driven by their need for precursors for macromolecular syntheses and their need for the energy to perform these polymerizations (Broda, 1975). The actual origin of life will not be discussed here and would be unprofitable since there is no really attractive theory to explain the first steps towards the evolution of a cell. The present range of possibilities is discussed by Miller and Orgel (1974), whose book serves as an excellent source of references for this chapter. It is their opinion, based on convincing arguments, that even very early biological entities must have had both nucleic acids and proteins; a triplet code was used and some intermediary system (like transfer-RNA) was capable of specific recognition of both amino acids and triplets. The amino acid could then be polymerized to form polypeptides

30

with a nucleic acid-determined composition. We think that the simplest self replicating "cell" would have contained, at a minimum, informational nucleic acids (DNA or RNA) which coded for a polymerase which would promote gene replication. It would probably also code for the transfer-RNA-like molecules. This cell would have been bounded by a membrane. Such a cell would require activated amino acids and nucleotides within the cell membrane.

Could the environment that existed on earth sometime between 4·6 and $3·2 \times 10^9$ years ago have provided these precursors? We shall look at the possibilities under the next heading.

2.1.1. Nutrients Synthesized Abiotically

When the earth was formed from cosmic dust, all the light components such as CH_4, NH_3 and H_2O would have been lost due to the weak gravitational field, had they not been combined chemically in non-volatile compounds. When the cosmic dust condensed there was a release of gravitational energy which caused the earth to become molten, possibly with a solid crust. The combined compounds were pyrolyzed, releasing CO, CH_4, H_2, N_2, NH_3, H_2S, and H_2O to the atmosphere. Possibly some CO_2 was also present. The relative proportions of these volatiles is not known as the equilibria are temperature-dependent. At lower temperatures (300°C) CH_4 and NH_3 would predominate, at higher temperatures (500°C) CO and N_2 are the stable species. Miller and Orgel (1974) calculate that the hydrogen pressure must have been between 10^{-4} and 10^{-2} atm. Numerous experiments have been performed on these reduced gas mixtures, subjecting them to electric discharge, light irradiation, heat, shock waves and variations in pH in order to study the abiological synthesis of organic molecules. It seems likely that the following compounds could have been formed abiotically $4·5 \times 10^9$ years ago. The amino acids with the exception of methionine and tryptophan may be synthesized abiotically without too much trouble. Acetate and propionate, but not the longer chain fatty acids, may be synthesized from methane. Porphyrins may be synthesized in small yield from formaldehyde and pyrrole, themselves produced from CH_4, NH_3 and H_2O. The purine adenine may be synthesized in good yield from refluxed HCN. Guianine also has been synthesized by a similar reaction. The pyrimidines, cytosine, uracil and thymine, may be synthesized in good yield from cyanoacetylene with cyanate. Sugars may be produced from formaldehyde, but not under conditions which favor their stability. The formation of nucleosides (organic base + ribose) is difficult abiotically, little is known regarding nucleotide synthesis. Amino acids vary greatly in their stability, the aliphatic amino acids (glycine, alanine, valine, isoleucine, leucine), proline and lysine are stable; histidine is susceptible to acid hydrolysis; arginine, aspartic acid, cerine, threonine, systeine, glutamine, asparagine and tryptophan are relatively unstable. Sugars are relatively unstable. Adenine and guanine are stable, the pyrimidine ring is stable but the pyrmimidine bases are easily hydrolyzed. The nucleosides are unstable.

Stability of these compounds is temperature-dependent, there would thus have been advantages if the oceans had been cool. Another advantage of cold oceans would have been the possibility of obtaining a concentration of solutes by freezing out the water. It would seem likely that life could have arisen only in a concentrated organic solution where the organic molecules would have had maximum stability. Evaporation, adsorption, differential solution and lipid barriers have been discussed as being potentially important in achieving locally high concentrations of organic compounds. If all the carbon on the surface of the earth ($3000 \, g \, cm^{-2}$) had been synthesized into organic compounds and dissolved in the oceans, the concentration would have been 1%. It is unlikely that abiotic synthesis was so efficient but concentrating processes could locally have produced quite concentrated solutions.

It is impossible to know whether abiotic synthesis continued after the first cell appeared but there can be little doubt that the atmosphere was changing over the first billion years (Section 2.2) due to the escape of hydrogen. Presumably the probability of abiotic synthesis decreased over this time and previously synthesized organic material disappeared either by inherent instability or by biological action. A summary of the possible substrates available to the primitive cell are illustrated in Table VII.

2.1.2. The First Prokaryotes

The properties of the most primitive organisms are assessed by attempting to define the simplest (most primitive) mechanisms whereby they could obtain cell materials and energy. This is a field that encourages speculation and flights of fancy and to some extent we have indulged in these luxuries. The nature of our proposal is, therefore, highly speculative but we think justified.

It is customary to imagine the most primitive bacterium as having a fermentative energy metabolism (like present day *Clostridium* species) taking up sugars from their rich organic environment, obtaining energy by a type of substrate level phosphorylation and polymerizing readily available monomers into nucleic acids and proteins (Hall, 1971; Margulis, 1970). This picture may be true but there are some disadvantages to it: (1) Sugars are difficult to synthesize abiotically and they are unstable, therefore, large concentrations are unlikely. (2) There was probably little nucleoside or nucleotide in the environment, for the synthesis of nucleic acids.

TABLE VII. Nutrients Available $> 3.5 \times 10^9$ Years Ago.

Inorganic:		
H	H_2:	$10^{-2} - 10^{-4}$ atm. Low solubility in ocean.
N	NH_3:	High solubility in ocean; photolyzed in atmosphere; trapped in clay minerals.
	N_2:	Increasing due to NH_3 photolysis.
	HCN:	Trace amounts.
S	H_2S:	Combined with Fe; high solubility in ocean; photolysis slight.
C	CH_4:	High initial concentration; low solubility in ocean; photolyzed in atmosphere.
	CO_2:	Increasing concentration; high solubility in ocean; possible carbonate formation.
	CO:	Little
Organic:		
Amino acids		High concentration.
Sugars		Little.
Formaldehyde		High concentration.
Purines and pyrimidines		Medium concentrations.
Nucleosides		Little.

An alternative to a fermentative *Clostridium* as the earliest prokaryote has been suggested by Yčas (1976). This is postulated to be a "palhirrotrophic" (tide-trophic) organism capable of gaining energy from the alternation between high and low external salinity, as might be found in an estuary. Estuarine sodium concentration can vary from zero at low tide to 460 millimolar at high tide. Yčas argues that increases in external sodium concentration could drive phosphorylations. If tidal flow is a prerequisite for the initiation of life, the number of extra-terrestrial situations in which life might have evolved would be very many fewer than is presently estimated, limited to those planets with a moon, oceans and a land mass. Palhirrotrophy is an attractive possibility as it requires only one component to generate phosphorylated compounds, unlike the next alternative to the clostridial type, a Gram-positive phototroph, capable of assimilating carbon compounds from the environment, which requires two components, a pigment and an ATPase (Fig. 9).

This bacterium might have had the following characteristics:

(1) It could have had a simple photosystem such as that found in present day *Halobacterium halobium*, consisting of a carotene-type pigment, bacteriorhodopsin. Such a Gram-positive bacterium is not known to exist now but carotenes occur in all known Gram-negative anaerobic phototrophs. Carotenes are also found in many Gram-positive bacteria, e.g., species of *Sarcina* and *Micrococcus*, where they are now considered to play a role in the preven-

tion of photo-oxidation (Mathews and Sistrom, 1959). This type of photo-phosphorylation is inherently simpler than substrate level phosphorylation since it requires only a photosensitive pigment to generate a proton gradient across the cell membrane, and an ATPase-like enzyme to perform the phos-phorylations (Fig. 9). There is, however, no geological evidence of iso-

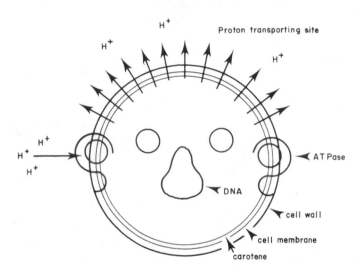

Fig. 9. A hypothetical primitive, Gram-positive phototroph generating a proton gradient through a carotene pigment and using this gradient to make ATP by means of an ATPase driven in reverse.

prenoids, degradation products of carotene-like compounds, before 3×10^9 years. ago.

(2) The original phototroph would have actively transported relatively dilute reduced carbon compounds from the environment (amino acids, sugars, and bases) and possibly phosphorylated them in the process of taking them into the cell. This would have had a very great advantage since they would be converted to a form in which they could readily undergo further conversions and polymerizations to form polynucleotides and polypeptides.

(3) The phototroph might very soon have depleted the environment of sugars which could then have been synthesized by steps, as for example, in the serine pathway (Fig. 7) utilizing sequentially serine, glycine, formaldehyde and finally CH_4 and CO_2. This would probably have been the order in which these compounds became limiting in the abiotic soup.

(4) The phototroph which acquired the capacity to photoassimilate CH_4 now have the necessary enzymatic pathway to utilize H_2 and CO_2 in a

respiratory process which it could use in the dark, generating CH_4. This set of characteristics would not be particularly strange or unique since the present day purple non-sulfur bacterium *Rhodopseudomonas gelantinosa* can photo-assimilate CH_4 (Wertlieb and Vishniac, 1967). The capacity to photo-assimilate reduced carbon compounds (ATP being generated through a proton gradient and coupled ATPase) while also possessing the capacity to respire reduced compounds in the dark is present in many purple non-sulfur bacteria (Evans, 1975). As we shall discuss later in the chapter, the O_2 respirers are probably derived from these purple non-sulfur phototrophs and in a parallel manner we propose that the H_2/CO_2 "respiring" methane producers could have arisen from the hypothetical carotene phototroph. It would be a rather satisfactory mechanism to account for the evolution of the methanogenic bacteria as there does not seem to be a clear line of evolution between substrate-level phosphorylation and electron/hydrogen transport phosphorylation as must exist in the methanogens. Very recent evidence, morover, suggests that the methanogens are not related to any other extant prokaryotes, suggesting an early and unique evolution (Fox *et al.*, 1977).†

Present day methanogens possess a correnoid methyl carrier which may not be directly involved in methane formation (Wolfe, 1971), but which has a tetrapyrrole structure. Tetrapyrroles can be synthesized abiotically (Miller and Orgel, 1974). Presumably it too was, in time, synthesized by the cell and this could have led to the evolution of the hemes and chlorophylls.

It would seem that this type of "respiratory" ATP-generating system might be mechanistically simpler than substrate-level phosphorylation but there is no real evidence that either the hypothetical phototroph or the methanogens came before the fermentative clostridia. There are certainly similarities between the methanogens and the clostridia in Gram-positivity, lack of cyto-chromes, occasional presence of correnoids, and the capacity of the latter to use CO_2 as an electron sink, via a correnoid and formaldehyde, generating acetate (Section 1.1.1.).

2.1.3. The Fermenters

Presumably if the clostridia were not the most primitive prokaryotes, they must have been close seconds in the process of evolution. They would have utilized the amino acids and the residues of abiotic synthesis in addition to fermenting the cell carbon of the primitive phototroph and the methane producers. Their products, in turn, the volatile fatty acids and alcohols,

---

† Most recently this has been supported further and evidence for some relationship to *Halobacterium* found [Magrum *et al.* (1978), *J. Mol. Evol.* **11**, 1–8 and Woese *et al.* (1978), *J. Mol. Evol.* **11**, 245–252] and the name "Archaebacteria" coined for these forms.

could have been converted largely to methane and carbon dioxide by the coupled reactions discussed earlier in Chapter 1. Reduced carbon compounds would have become even more limiting and methane would have been lost from the environment due to its poor solubility in water and its photolysis in the atmosphere. Presumably some ozone must have been in the atmosphere at this time, due to the photolysis of water, and this would have screened out some lethal UV irradiation. The stage was set for the emergence of the chlorophyll-containing phototrophs.

2.1.4. The Chlorophyll Phototrophs

Present day phototrophs all have the Calvin cycle for the reduction of carbon dioxide and all contain some form of chlorophyll, a magnesium porphyrin. They are all Gram-negative and thus have a basic dissimilarity from the primitive bacteria discussed. The photo-assimilators of carbon compounds may have existed at this time (purple non-sulfur bacteria) assimilating methane and the products of primitive metabolism but equally likely was the prior emergence of the sulfur phototrophs. A later emergence of the latter is, however, suggested by the fact that the ferredoxin of the green sulfur bacteria is probably intermediate between the primitive clostridial type and that of the more evolved *Desulfovibrio* species (Hall *et al.*, 1971). Reduced sulfur compounds might have been more available than reduced carbon compounds at that time as electron donors for carbon dioxide reduction, favoring the emergence of the sulfur phototrophs.

Broda (1975a) has reviewed the available evidence and concluded that the purple non-sulfur bacteria were the earliest, partly because of their photo-organotrophy. Peck (1974), on the other hand, believes that photo-driven oxidation of sulfide may have had very early beginnings and that the primitiveness of the purple sulfur bacteria is indicated by their very low guanosine and cytosine content (Pfennig, 1967). The oxidized sulfur products certainly became available at an early stage and were able to participate in the oxidation of hydrogen, lactate, acetate, etc. (CHO/SO_4^{2-}, see Section 1.1.1.) by newly emerged species of *Desulfovibrio* and *Desulfotomaculum*. These would have originated from the sulfur phototrophs (Peck, 1974) or possibly from the purple non-sulfur phototrophs (Broda, 1975a). Presumably any bacteria capable of sulfate reduction/methane oxidation may have arisen at this time too. A complete carbon cycle involving alternate oxidation/reduction of S was thus possible (Levy *et al.*, 1973; Peck, 1974). Apparently no such cycle involving N had originated at this time. Ammonia is not known to act as an electron donor in any bacterial photosystem and consequently was not oxidized, neither is it known to be oxidized anoxically by any chemotrophic system (Broda, 1977). It is thus very conservative in anaerobic environments and was probably oxidized in primitive times only by photolysis. It would seem, however, that ammonia became limiting in early biological history either due to photolysis, biological assimilation or by being trapped in clay minerals. The evidence for this lies in the prevalence of nitrogen-fixation in so many primitive anaerobic bacterial types, as discussed in Section 1.2.2. Other possible roles for nitrogenase have been suggested (Silver and Post-gate, 1973), and Postgate (1974) has pointed out that the easy transference of the nitrogenase genes might have resulted in primitive bacteria acquiring the property at a much later time. The appearance of heterocysts which have a unique function in nitrogen-fixation, at an early time, would indicate an early origin. Heterocysts are seen in fossil blue-green bacteria (Section 2.2.2.).

2.1.5. The Oxygen Evolving Phototrophs

In addition to the primitive sulfur phototrophs it has recently been shown that some blue-green bacteria can utilize reduced sulfur compounds as electron donors for carbon dioxide reduction (Cohen et. al., 1975; Garlick et. al., 1977). This is a significant discovery since it is a bridge between anaerobic and the oxygen-evolving photo-processes. It is not known from which of the primitive phototrophs the blue-green bacteria evolved; the information quoted would indicate a relationship to the sulfur photosynthetic bacteria. Olson (1970) considers that the blue-greens and the anaerobic phototrophs may have arisen from a common photo-assimilatory ancestor. One of the main reasons for this conclusion is that chlorophyll a (blue-greens) is probably an intermediate on the pathway to the biosynthesis of bacteriochlorophyll a, suggesting that the common ancestor had a chlorophyll a structure and that bacteriochlorophyll evolved from it.

The possession of a photosystem capable of utilizing water rather than reduced C and S compounds was an enormous advantage as these compounds were in short supply even if they were being regenerated to some extent by e.g., Desulfovibrio species. Their transport from the site of reduction to the photosynthetic bacteria would have been limited by diffusion. Water, however, was available everywhere. The presence of biologically generated oxygen in the atmosphere had very profound effects, the most obvious being the evolution of O_2 respiration.

2.1.6. The Oxygen Respirers

Initially oxygen was probably toxic to the flavin-containing anaerobes but peroxide-detoxifying systems evolved (peroxidases and superoxide dismutases) which allowed oxygen to be used at the end of a cytochrome electron transport chain. This enabled bacteria to obtain maximum ATP yield from reduced substrates which in turn gave a greater efficiency in substrate assimilation (see O_2-oxidation in Section 1.1.1.). It resulted in an extension of the S-cycle in providing a non-photomechanism for S-oxidations. It also resulted in the emergence of the nitrifying bacteria which produced nitrite and nitrate. These in turn led to the process of denitrification in which oxidized nitrogen compounds, particularly nitrate, act as an electron sink, and are converted to dinitrogen. This added a completely new dimension to N-cycling and presumably increased the necessity for nitrogen-fixation. It also led to the possibility of hydrogen oxidation by oxygen, thus completing a biological water cycle. Oxygen was also involved in the oxidation of reduced iron compounds.

Probably eukaryotic evolution was dependent on the emergence of oxygen respiration. There is good evidence that mitochondria, the energy transducers of eukaryotic cells, are derived from endosymbiotic prokaryotic respirers (Margulis, 1970). The prokaryotic cells have evolved many other less integrated symbiotic relationships of great biological importance, which will be discussed in Chapter 8.

The presence of oxygen in the atmosphere made possible the oxidation of methane by oxygen. The ancestors of the diverse types of methane-oxidizing bacteria are unknown but their carbon metabolism and "autotrophy" via formaldehyde is similar to the CH_4/SO_4^{2-} and H_2/CO_2 bacteria, possibly indicating a line of descent. The evolution of the main groups of CHO/O_2 oxidizers is beyond the scope of this book. It seems very probable that they have evolved from the purple non-sulfur bacteria, which already contained the main electron transport chains involving flavins and cytochromes necessary for the efficient transfer of electrons to oxygen.

Not all evolution has been towards increased complexity and versatility since it is possible that some respiring bacteria may have lost this capacity and reverted to a fermentative metabolism, e.g., the lactic acid and propionic bacteria (Broda, 1975a). Similarly some blue-green bacteria may have lost their photosynthetic capacity and become colorless sulfur bacteria or chemoheterotrophs just as the purple bacteria probably became respirers. Margulis (1970, 1972) discusses these systems in more detail.

The comparison of the amino acid sequences of ferredoxins, low potential electron acceptors involved with N_2-fixation, CO_2 reduction and H_2 metabolism, has been mentioned as a tool for the evaluation of evolutionary distances (Hall et al., 1973). Cytochrome c can be used in a similar manner but as it does not occur in bacteria more primitive than Rhodospirillum rubrum (McLaughlin and Dayhoff, 1970) and Desulfovibrio species, it has been more useful in elucidating eukaryotic evolution.

The general mechanism for the evolution of enzymatic pathways has been discussed by Horowitz (1965) who has suggested that as nutrients became depleted in the primitive environment, the bacteria evolved new enzymes capable of utilizing some other related component. This would have been converted to the required compound. Miller and Orgel (1974) extend the argument and consider that intermediary metabolism may be a reversal of the pathways of degradation of biochemicals in the external soup. Thus the first degradation step could be reversed by a single synthetic step, possibly mediated by the enzyme which already had an affinity for the product. Once the first degradation product had been exhausted, the cell would have been forced to use the next degradation product by a modification of an existing enzyme, presumably again one which had some affinity for it, namely the enzyme which recognized the product. In this way, complex synthetic pathways could have evolved in discrete steps. The synthetic pathways for aromatic amino acids, nucleosides and long chain fatty acids are similar in all forms of life and presumably evolved early and have been conserved.

Present day evidence for the evolution of new enzyme functions can be demonstrated in the laboratory and is thought to occur by these general principles (Clark, 1974).

2.2. GEOLOGICAL EVIDENCE

The evolutionary considerations given so far are based on metabolic pathways or other biochemical features of extant prokaryotes. Within the last two decades new direct evidence with respect to the early life on earth and the prevailing environmental conditions has appeared. However, as shown below there are still large gaps in our knowledge and problems of interpretation are considerable. In the following we will review some of the palaeobiological evidence which is relevant to the evolution of element cycling.

2.2.1. The Time Scale

The age of the earth is now believed to be 4.6×10^9 years. This figure is based on three different estimates. The moon and meteorites are believed to have formed simultaneously with the earth. Dating of the oldest moon rocks collected on the Apollo missions and of meteorites (using the radio-active decay rates of uranium, rubidium and potassium isotopes) both yield an age of 4.6×10^9 years. Using the decay rate of ^{238}U to ^{206}Pb and the ratio ^{206}Pb to ^{204}Pb on earth also gives an age of about 4.6×10^9 years (Miller and Orgel, 1974). The oldest dated rock from earth (Greenland) is estimated to be 3.75×10^9 years (Cloud, 1974); from the first 0.9×10^9 years of the earth's history we have no direct evidence.

Geological time is usually divided into two periods; the *Precambrium* and the *Phanerozoicum*. The latter comprises the last 570×10^6 years and thus only about 12% of the history of the earth. The beginning of the Phanerozoic period (*Cambrian*) is characterized by a sudden richness in the record of fossil metazoans and in the following (*Ordovician*) period all extant invertebrate phyla are represented. In the *Silurian* period ($400–420 \times 10^6$ years ago) aquatic vertebrates appear and in this and in the following period (the *Devonian*) the terrestrial world was colonized by animals and vascular plants.

With the exception of few, generally poorly preserved invertebrate fossils from the period around the Cambrian-Precambrian border, the vast Precambrian period of about 4×10^9 years seemed until the mid 1950's only indirectly and ambiguously to give evidence of life. Among the new evidence is the finding of Precambrian microfossils which establish the Precambrium as a world of prokaryotes and which provides us with some time markers for important events in the evolution of the biosphere.

2.2.2. Precambrian Fossils

All continents have nuclei of Precambrian rock shields (e.g. the Fenno-Scandian Shield and the Canadian Shield). Unfortunately these consist mainly of magmatic or strongly metamorphosed sedimentary rocks. However, in most continents Precambrian sedimentary horizons consisting of cherts, shales, limestones, or dolomites have been found which contain microfossils. In some cases these fossils are remarkably well preserved due to the rapid impregnation with amorphous silica, after the death of the organisms. Precambrian fossils have been reviewed by Swain (1969), Schopf (1970, 1974), and by Barghoorn (1971).

The oldest deposit with objects which perhaps could be interpreted as microfossils consists of black cherts in the *Fig Tree Formation* in Swaziland; they have been dated to $3 \cdot 1 \times 10^9$ years. The fossils consist of about $0 \cdot 5 \mu m$ long bacteria-like rods and some of granular, spheroidal cells with coalified organic matter. The latter have been compared to coccoid blue-green bacteria. In fact, some palaeochemical evidence seems to indicate the presence of photosynthesis during the deposition of the *Fig Tree Formation* as well as in the underlying *Onverwacht Group* which is estimated to be $3 \cdot 2 \times 10^9$ years old. As discussed below, this photosynthesis need not have been oxygenic. The Onverwacht deposits have also yielded some irregular, carboniferous microbodies; if these are remains of cells, they represent the oldest direct evidence of life on earth so far found. There are, however, reasons to suspect that the structures found in the Onverwacht deposits and perhaps even those of the *Fig Tree Formations* are of abiological nature and not bacterial fossils (Cloud, 1976; Schopf, 1976).

Evidence of filamentous prokaryotes are thread-like microfossils from the *Soudan Iron Formation* (Minnesota) which are about $2 \cdot 7 \times 10^9$ years old and the stromatolites from the *Bulawayan Limestone* (Rhodesia) which are $2 \cdot 8 \times 10^9$ years old. Stomatolites are laminated calcarious or silicious structures which are well known from the younger Precambrian and Phanerozoic geological record. Their modern counterparts are layered communities of filamentous blue-greens, especially oscillatorians, which may take part in the limestone deposition in extant coral reefs. The stromatolites from the *Bulawayan Limestone* are thus taken as evidence for the presence of filamentous blue-greens $2 \cdot 8 \times 10^9$ years ago (Schopf, 1970).

When interpreting these findings of microfossils from early Precambrian ($> 2 \cdot 5 \times 10^9$ years) many reservations must be made. It is believed that at least some of the microfossils are not abiological artifacts and that they are syngenetical with the formation, an assumption which is supported by the palaeochemical evidence. Even so, very little can be inferred about metabolic activity on the basis of morphology. Prokaryotes have not been very inventive with respect to morphology and there are many examples of extant bacteria which look quite similar but are physiologically very different (Fig. 10). Recently, filamentous bacteria with bacterio-chlorophylls and anoxygenic photosynthesis have been found in hot springs. They are morphologically difficult to distinguish from blue-greens and they form siliceous stromatolites (Pierson and Castenholz, 1971; Walter *et al.*, 1972). The fact that extant blue-greens may utilize H_2S in anoxygenic photosynthesis (Cohen *et al.*, 1975) also calls for caution when interpreting prokaryote fossils. A critical discussion on the Early Precambrian microfossils is given by Cloud (1976).

Similar reservations apply when the much richer findings from the Middle Precambrian ($2 \cdot 5 - 1 \cdot 7 \times 10^9$ years) are discussed. However, the good preservation and many structured details of some of these fossils do seem to establish the presence of "modern" groups of blue-greens. Micro-fossils from this period have been found in many Precambrian sedimentary rocks from several localities (South Africa, Greenland, Canada, Australia). They reveal communities of filamentous and coccoid blue-greens and other bacteria. Most remarkable are the findings from the Gun Flint Chert in Ontario. One silicious stromatolitic horizon of up to 60 cm thickness and with a wide horizontal distribution has yielded a three-dimensionally preserved flora of bacteria some of which have unusual shapes and others which have been compared to modern bacterial genera like *Crenothrix*, *Metallogenium*, and *Sphaerotilus*, coccoid blue-greens and filamentous forms belonging to the Oscillatoriaceae and the Nostocaceae. The filaments of the latter show swellings which can only be interpreted as heterocysts. Since heterocysts in

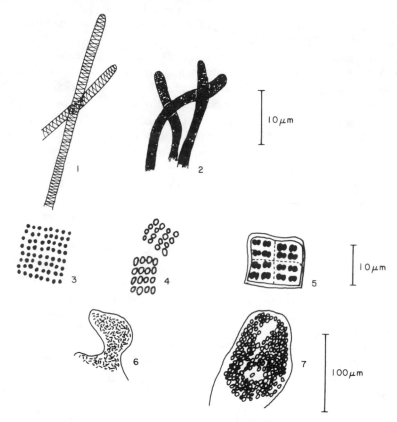

Fig. 10. Examples showing that bacteria with similar morphologies may have very different physiological properties. 1. *Oscillatoria* (oxygenic photoautotroph); 2. *Beggiatoa* (chemolithotrophic sulfide oxidizer); 3.*Lampropedia* (chemoheterotroph); 4. *Thiopedia* (photosynthetic sulfurbacterium); *Eucapsis* (oxygenic photoautotroph); 6. *Zoogloea* (chemoheterotroph); 7. *Microcystis* (oxygenic photoautotroph).

extant blue-greens are specialized, non-photosynthetic cells used for N_2-fixation in aerobic environments, this finding would indicate the presence of N_2-fixation as well as aerobic environments in Gunflint times.

Late Precambrian ($1\cdot 7 - 0\cdot 57 \times 10^9$ years) has left well preserved microfossils in many localities (North America, Scandinavia, Australia, Asia). The best preserved and studied flora is that of the *Bitter Spring Formation* in Australia which is dated to approximately $0\cdot 9 \times 10^9$ years. A diverse flora of bacteria including blue-greens has been revealed. In addition to the groups found in the Gun Flint, Rivulariaceae and colonial coccoid blue-greens have now appeared. Among the blue-greens many can be assigned to extant genera. Even more exciting is the first convincing evidence of eukaryote $8\mu m$ spherical cells of which some have been found in stages of mitosis. Somewhat more daringly, other fossils from the Bitter Spring have been interpreted as red algae, fungi and dinoflagellates. Microfossils from the somewhat older Beck Spring Dolomite (California) have also been interpreted as eukaryote cells; the evidence for this, however, is not very strong.

To complete this short account on Precambrian fossils, the oldest known remains of metazoans derive from the Australian *Pound Quartzite* signalling the rapid diversification and evolution of metazoans and metaphytes in the Cambrian and later periods. Some of the important localities for Precambrian fossils are listed in Fig. 11.

The fossil records establish the greater part of the Precambrian as the era of prokaryotes. Together with the palaeochemical data discussed below, the fossils suggest that oxygenic photosynthesis is at least 2×10^9 years old (Gun Flint) or perhaps even 3.2×10^9 years old (Onverwacht) and that anoxygenic photosynthesis may at least be that old. Thus, during the preceding 1.4×10^9 years, the origin of life, the evolution of the prokaryotic cell, and the major part of the evolution of metabolic pathways, as outlined in Section 2.1 and Section 2.2.4, must have taken place. Unfortunately we have no evidence of life older than that of the Onverwacht Formation.

2.2.3. Chemical Fossils

Instead of, or in addition to recognizable fossils, sedimentary rocks may contain organic matter of biogenic origin which yields information about the life that synthesized it. Biogenic sediments are of course well known from Phanerozoic deposits (e.g., lignite, mineral oil) and such deposits may contain more or less denatured, specific compounds yielding specific taxonomical or palaeoecological information. During the last 15 years conserable attention has been given to organic compounds of Precambrian deposits (e.g., Eglinton and Calvin, 1967; Eglinton and Murphy, 1969; Schopf, 1970; Jackson and Moore, 1976; Kvenvolden, 1976).

Several different groups of organic compounds have been studied; in many cases, however, results must be considered with reservation (e.g., amino acids). In general, any specific group of organics is present in extremely small concentrations in the Precambrian rocks. When interpreting reports on the presence of a compound, three problems have to be taken into consideration. Due to the small amounts present in the rocks and the omnipresence of organic compounds today there is a real risk of contamination during sampling and treatment of rock samples. This problem is believed to be solved in general, by using the utmost care during the laboratory procedures. A more serious objection is whether the studied compounds are syngenetic with the rock or whether they entered it at a later stage through cracks or by

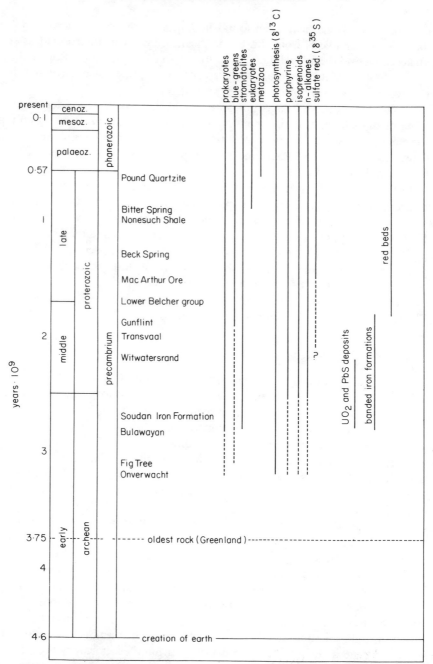

Fig. 11. The age of the most important geological formations containing Precambrian fossils and various palaeochemical evidence of Precambrian life and atmospheric O_2 content.

diffusion. This objection has in fact brought some of the reported findings in doubt as evidence for Precambrian life. If the studied compounds show an isotopic ratio between ^{12}C and ^{13}C (see below) which differs from that of the bulk of unextractable organic matter in the rock, this may be taken as evidence that the compound is not syngenetic with the rock but may have entered it later, or be due to contamination. Smith *et al.* (1970) have seriously questioned the syngenetic origin of extractable organics from some Precambrian deposits, including some of the findings discussed below. The last problem is whether the compound in question is of biotic origin. For the compounds discussed below this seems likely. Thus in the case of e.g., alkanes, normal alkanes dominate over a large number of isomers which are thermodynamically equally probable, had the alkanes been synthesized abiotically (Schopf, 1970).

Three types of organic material extracted from Early and Middle Precambrium are generally accepted as bona fide examples of chemical fossils which give some information on the biology of the organism which formed them: alkanes, porphyrins and the isoprenoid hydrocarbons, pristane and phytane. Porphyrins are components of pigments (e.g., chlorophylls) and pristane and phytane are believed to be the remains of the phytol side chain of chlorophylls. These compounds have all been demonstrated in (among other deposits) the Fig· Tree and Gun Flint Formations. If they are of syngenetical origin, then they support the assumption that prokaryote photosynthesis (oxygenic or anoxygenic) took place $\sim 3\cdot1 \times 10^9$ years ago.

Another method for detecting fossil photosynthesis is based on the fact that the stable isotopes ^{12}C and ^{13}C undergo a fractionation during photosynthesis so that organic material is enriched in ^{12}C relative to the isotope ratio of CO_2 (Park and Epstein, 1960). The isotope ratio is usually expressed as:

$$\delta^{13}C = [(^{13}C/^{12}C \text{ sample})/(^{13}C/^{12}C \text{ standard}) - 1] \times 1000\%_0$$

and where the carbonate of a belemnite fossil is used as a standard. An enrichment of ^{12}C in organic carbon relative to inorganic carbon has in fact been found in the Onverwacht and the Fig Tree Formations, again indicating that photoautotrophy occurred already $3\cdot2 \times 10^9$ years ago (Oehler and Schopf, 1972; Kvenvolden, 1976). However, as with other evidence pertaining to early life, interpretations of stable isotope ratios are still somewhat uncertain and open to discussion (Cloud, 1976).

It would be of considerable interest to establish the antiquity of dissimilatory sulfate reduction since this process may well have been the first kind of respiration to exist. Trudinger *et al.* (1972) give various indirect evidence that some Late Precambrian sulfide ores are of biogenic origin. An isotope ratio method, similar to the $^{12}C/^{13}C$ method is in principle available for determining whether sedimentary sulfide is biogenic. Sulfate reducers show a slight kinetic discrimination between the two natural sulfur isotopes so that H_2S formed by sulfate reduction is enriched in ^{32}S relative to the sulfate. The isotope ratio of a sulfur species is expressed as:

$$\delta^{34}S = [(^{34}S/^{32}S \text{ sample})/(^{34}S/^{32}S \text{ standard}) - 1] \times 1000\text{‰},$$

and where the standard used is troilite from a meteorite which is assumed to represent the average terrestrial isotope ratio (e.g., Kemp and Thode, 1968; Peck, 1974).

The method has been used among other evidence, for demonstrating the biogenic nature of more recent, i.e., Phanerozoic sulfur deposits and sulfide ores (e.g., Ivanov, 1968; see also Section 6.2.7). Ault and Kulp (1959) report enrichment of light sulfur from sulfides in rocks about 2×10^9 years old and interpret these as a sign of biogenic sulfide formation whereas older rocks ($\sim 3 \times 10^9$ years) yield average crust values. However, the isotope ratio method gives variable results which may be difficult to interpret. Thus the isotope ratio of oceanic sulfate has changed through geological time, not only due to biological activity but also due to varying rates of evaporite formation (Rees, 1970) and unless the isotope ratios of more, syngenetically deposited, sulfur species are known, information from isotope ratios is limited. Also, a number of mechanisms involved in the isotope fractionation due to the microbial sulfur cycle are incompletely understood. There is, therefore, not yet any conclusive geological evidence with regards to the origin of sulfate reduction.

2.2.4. The Evolution of the Biosphere and Microbial Element Cycling

This section will finally summarize the different ideas and evidence discussed in this chapter in context with current knowledge and conjectures on the chemical evolution of the biosphere in order to arrive at some understanding of the evolution of element cycling. This section will only treat qualitative aspects. A more quantitative description of the present biosphere and its global element cycles will be given in Chapter 9.

It is generally accepted that the present atmosphere is not in a chemical equilibrium and that the maintenance of a dynamic steady state inequilibrium is mainly due to biological activity. Thus, an atmosphere containing both O_2 and N_2 is thermodynamically unstable, since, for

$$N_2 + 5/2O_2 + H_2O \text{ (liq)} = 2HNO_3 \text{ (aq)},$$

$K = 0.04$ at $25°C$ (Sillen, 1966; Miller and Orgel, 1974). Thus were it not for microbial denitrification, atmospheric N_2 would, due to electrical discharges, eventually oxidize to transform the oceans into a 0.01 M solution of nitric acid. Similarly the atmospheric oxygen level is maintained by photosynthesis; otherwise it would decrease due to the oxidation of reduced carbon, sulfides and other components in soil and rocks. While it is difficult to reach a full quantitative understanding of the present chemical composition of the biosphere, considerations about the development through geological time must be highly speculative on the basis of current evidence.

As already discussed, it is believed that the primary atmosphere of the earth was lost very early in the history of the planet and thus a great part of its original content of hydrogen disappeared. This is concluded on the basis of the very low concentration of noble gases, not derived from radioactive decay, in the present atmosphere. The secondary atmosphere of the earth was formed by outgassing. Since the gases from contemporary volcanos do not derive from juvenile material, we have no way of knowing the composition of the gases which originally formed the atmosphere, but we may assume that it contained CO_2, CO, CH_4, H_2, NH_3, H_2O, and perhaps H_2S. Even if we knew the elemental composition, considerations on chemical equilibria would be insufficient for calculating the quantitative occurrences of different compounds without taking kinetic aspects into account. This would be true even for the abiotic earth, due to processes such as photolysis of molecules in the upper atmosphere, loss of hydrogen atoms to space, and chemical reactions with tectonically exposed minerals.

It is generally accepted that the early atmosphere had a very low oxygen pressure which increased through geological time at least in part, due to the appearance of oxygenic photosynthesis. There is little evidence, however, to support the belief that the early atmosphere was strongly reducing, i.e., dominated by CH_4 rather than CO_2. The only valid argument is, in fact, that the abiotic synthesis and stability of organic compounds are only possible in a reducing environment as discussed in Section 2.1. Without accepting the Haldane-Oparin ideas on the origin of life in an "organic broth", it is (even more) difficult to understand how life originated and how the first organisms made their living.

Irrespective of how reduced the early atmosphere was, it is agreed upon that it slowly tended to become more oxidized even in the absence of organisms. Water, CH_4, and NH_3 are photolyzed in the upper atmosphere and the light hydrogen atoms may escape from the gravitational field of the earth. This process is probably self-limiting to some extent since the O_3 molecules formed will create a UV-shield and block further photolysis (the "Urey effect"). The O_2 (and the more reactive species O and O_3) would combine with various reduced compounds on the surface of the earth (e.g., ferrous iron and sulfides) so that the actual O_2 level would have been low in the pre-photosynthetic atmosphere (Berkner and Marshall, 1967; Cloud and Gibor, 1970). The level at which atmospheric oxygen would equilibriate under these conditions, however, is controversial. After the advent of oxygenic photosynthesis, O_2 was produced, but the Urey effect, the abiotic oxidation of soils and minerals, and respiration could have kept the level low until abiotic oxidation had slowed down and respiration of organic material had lagged behind, so that atmospheric oxygen could build up. According to one estimate (Berkner and Marshall, 1967) the O_2 level was kept below 0·1 % of the present level until Late Precambrian. These authors suggest that the level reached about 1 % at the beginning of the Cambrian, reached the present level during the middle of the Palaeozoic period, overshot during Late Carboniferous to reach the present level again through some oscillations. According to another estimate (Rutten, 1970), the O_2 level was controlled by the *Pasteur point* after the advent of oxygenic photosynthesis and respiration. The Pasteur point is the minimum oxygen tension at which oxidative respiration takes place. In extant aerobes this is about 1 % of the present oxygen level (Fig. 12). Broecker (1970), based on the constancy of $\delta\,^{13}C$ values of carbonates, argues that the oxygen content of the atmosphere cannot have changed significantly within Phanerozoic times.

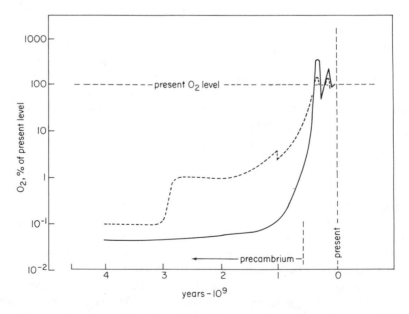

Fig. 12. The evolution of the atmospheric O_2 content. Interrupted line according to Rutten (1970), uninterrupted line according to Berkner and Marshall (1967).

There is some direct evidence to show that the present oxygen-containing atmosphere evolved from an anoxic one. Minerals deriving from Early Precambrian (e.g., sulfur, iron) occur mainly in the reduced state (or if not, they are believed to have oxidized secondarily during a later geological period). The banded iron formations ($2 \cdot 7 - 1 \cdot 8 \times 10^9$ years) consist of alternating layers of Fe_3O_4 and iron-poor silica. The formation is usually explained as the effect of precipitation of ferric iron in low oxygen concentrations from dissolved ferrous iron originating from weathering rocks. Another interpretation is that during this period the seas were essentially anoxic and reducing but with an oxidized epilimnion (much like the Black Sea in present time). By oscillations in the vertical position of the oxic-anoxic boundary layer the banded structure may be explained (Degens and Stoffers, 1976).

The banded iron formations are followed by the deposition of red beds, i.e., deposits with the more oxidized Fe_2O_3 corresponding to a higher O_2 pressure in the atmosphere.

The recent findings on the environment on Mars may contribute somewhat to the understanding of terrestrial development. The Martian atmosphere consists mainly of CO_2 (95%), N_2 (2–3%) but only $0 \cdot 1$–$0 \cdot 4$% O_2 and yet the planet has an oxidized surface (Owen and Biemann, 1976). Thus, it shows that an inner planet may, presumably abiotically, develop an oxidized atmosphere but with a very low oxygen concentration and also that the surface may oxidize under a very low oxygen tension.

Large deposits of UO_2 and PbS which are about $2 \cdot 5 \times 10^9$ years old indicate the presence of widespread anoxic conditions since both minerals oxidize at low oxygen tensions. In the case of some UO_2 deposits it has been suggested that they were formed by weathering and solution of U-containing rocks in an oxidizing atmosphere; the dissolved UO_3 was then brought by rivers to estuaries or basins which were still reducing and anoxic so that the metal could precipitate as UO_2 (Cowan, 1976). The fact that many prokaryotes, and among them many which are believed to be primitive with respect to metabolism, are obligatory anaerobes also indicates that the biosphere was once anoxic. In eukaryotes, aerobicity is a fundamental property and anaerobicity is only found among a few specialized (mainly parasitic) forms. The occurrence of eukaryotes about 10^9 years ago indicates the availability of oxygen in at least some environments.

If we assume that the first organisms in Early Precambrian times made their living by fermentation of abiotically synthesized organics, by some kind of photoheterotrophy or by membrane potential phosphorylation, as discussed in Section 2.1.2, then there was, in one sense, only one element cycle, viz. the carbon cycle. The organisms would of course assimilate and excrete various other elements but C was the only element to enter the energy yielding reduction-oxidation processes. As previously discussed, photosynthetic sulfur

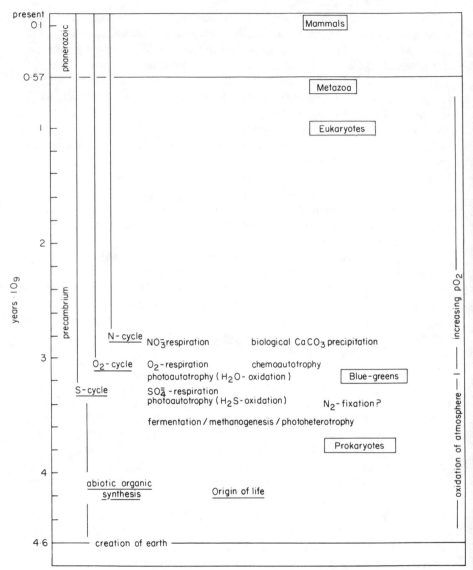

Fig. 13. Important steps in the evolution of prokaryote metabolic types and of the major element cycles.

bacteria may have been among the first autotrophs to develop from photo-heterotrophic ancestors. Once these forms started the photosynthetic oxidation of reduced sulfur compounds to SO_4^{2-} the world was set for the first kind of respiration, using SO_4^{2-} as electron acceptor. A light-driven sulfur cycle was thus established, making organisms independent of abiotically synthesized organics. Unfortunately there is no geological evidence to verify or falsify such a line of events.

The next step in the evolution of element cycling could have been oxygenic photosynthesis followed by aerobic respiration. These events may be reflected in the banded iron formations and in fossils of blue-greens in the Middle or perhaps Early Precambrian deposits. The advent of a light-driven oxygen cycle and thus the availability of O_2 gave possibilities for the evolution of chemoautotrophs, oxidizing substrates such as Fe^{2+}, NH_3 and H_2S.

The appearance of oxygen also made the closing of the nitrogen cycle possible. The presence of NO_3^- in an anoxic world is not likely and denitrification is therefore probably a more recent invention than is aerobic respiration. This conjecture is not contradicted by the taxonomic distribution of denitrifiers in extant bacteria (see Section 5.3). A "modern" nitrogen cycle was, therefore, probably the last major element cycle to evolve. An alternative to this speculation is to claim the existence of an extinct (or undiscovered) photosynthetic form which used reduced nitrogen compounds as electron donor. Such an organism is thermodynamically possible and has been discussed by Olsen (1970) as a possible link between anoxygenic and oxygenic photosynthesis. Other aspects of the evolution of the nitrogen cycle remain problematical. Nitrogen fixation is believed to be a primitive feature found in many extant forms of prokaryotes with primitive metabolic pathways. Also nitrogen fixation seemed to have taken place in the blue-green flora in Gun Flint time (Section 2.2.2). Even though the early atmosphere was reducing, it is possible that NH_3 only occurred in small amounts due to photolysis in the upper atmosphere and due to the high reactivity of NH_3. It is, therefore, well possible that N_2 was the dominating form of atmospheric nitrogen and that nitrogen fixation was adaptive in the early biosphere. On the other hand, it is hard to see what prevented reduced nitrogen from accumulating in the biosphere (besides the loss through photolysis in the upper atmosphere of NH_3) as long as O_2 was absent.

The course of events described above are summarized in Fig. 13. Here it is assumed that oxygenic photosynthesis evolved around Fig Tree time. It might also have taken place 0.5×10^9 years later or even earlier than Fig Tree. The main conclusion, however, is that by Late Precambrian many of the fundamental properties of the extant biosphere, including the major element cycles driven by biological processes, had already evolved. Thus, it is mainly prokaryotes and prokaryotic metabolic diversity which have shaped the chemical environment in which eukaryotes originated, diversified and evolved.

CHAPTER 3

Bacteria in Detritus Food Chains

The microbial element cycles are mainly driven by the chemical energy of dead organic matter; even the electron donors of phototrophic eubacteria derive from the incompletely mineralized dead organic matter. Although parasitic and predatory (e.g., *Bdellovibrio*) bacteria exist, their role in the carbon cycle of ecosystems is trivial. The most important ecological role of bacteria is, therefore, the mineralization of dead organic matter and through this process element cycles are driven. In this chapter we will discuss the initial stages of mineralization processes and some of their quantitative aspects.

In most ecosystems a large or even dominating fraction of the organic production is not consumed by herbivore or carnivore animals but is added to a pool of dead organic matter, "detritus". This material is then predominantly broken down by microorganisms. These have a number of features which explain their dominating role as primary decomposers of detritus. Bacteria can hydrolyze a diverse assemblage of organic compounds, some of which are totally undigestible for animals. (This observation is valid for bacteria as a group; single bacterial strains only utilize a more or less restricted number of substrates.) Furthermore bacteria have the ability to efficiently take up and utilize substrates (carbon sources as well as mineral nutrients) in dilute solutions. The small size of bacteria in conjunction with surface-bound hydrolytic enzymes makes possible a close contact with solid substrates and minimizes loss of the products. The last two mentioned properties of bacteria are due to their small size and consequently a large surface to volume ratio and also to efficient specific uptake systems. Finally, bacteria can mineralize organic substrates efficiently under anaerobic conditions.

Since the term detritus is ambiguous in the literature, it will be useful to define the term as used here. By detritus we mean *the organic carbon lost by non-predatory processes from any trophic level (this includes egestion, excretion, secretion, etc.)* and by a *detritus food chain* is understood *any route by which chemical energy contained within detrital organic carbon becomes available to the biota* (Wetzel *et al.*, 1972). Note that this definition excludes the living component of, e.g., particulate detritus, and that it includes dissolved organics.

52

It also recognizes that the energy transfer in a detritus food chain may involve inorganic compounds (e.g. H_2S, NH_3).

Table VIII shows estimates of the fraction of the primary production which is consumed by herbivores in different plant communities. It is seen that it is relatively small and tends to decrease with increasing sizes of dominating primary producers. The remaining part of the production enters the pool of detritus as particulate matter or as dissolved leachates from living or dead plant material.

TABLE VIII. Fraction of Primary Production
Consumed by Herbivores.

Plant community	% production consumed[a]
Phytoplankton	60–90
Grass lands	12–45
Kelp beds	~10
Spartina marshes	~7
Mangroves	~5
Deciduous forests	1·5–5

[a] Compiled from Woodwell (1970), Wiegert and Owen (1971) and Fenchel and Jørgensen (1977).

The reason why such a relatively small portion of the primary production of most plant communities is consumed by herbivores (i.e. "why the world is green") has been discussed by several authors (e.g., Hairston *et al.*, 1960; Wiegert and Owen, 1971; Fenchel and Jørgensen, 1977). The functional explanation is that very few animals (excluding forms with symbiotic relationships with microorganisms) are capable of utilizing the tissue of higher plants since animals lack digestive enzymes which attack structural plant compounds such as cellulose and lignin. Also, plants often have a number of anti-predatory devices (toxins, aromatic resins, thorns). Finally most plant tissue is poor in mineral nutrients (in particular, nitrogen and phosphorus). Russell-Hunter (1970) generalized that—with the exception of ruminant mammals—animals require a diet with a C/N ratio of less than about 17 and the tissue of plants generally has considerably higher C/N values. From an evolutionary point of view, these features of plants have been adaptive mainly by limiting herbivory and thus allowing the evolution of large plants with a large biomass to productivity ratio.

There are contributions to the pool of detritus other than dead plant tissue. Various types of particulate matter derive from crustacean exuviae, pollen, etc., and large amounts of dissolved organics are due to the excreta of plants, animals and microorganisms.

On a global scale the pool of dead organic material is nearly in a steady state and only an infinitesimal fraction (as compared to the biological productivity) is stored in sediments as organic carbon (including fossil fuels) or as reduced sulfur. This storage is significant only over geological time. By far the greatest part of the production is mineralized, albeit with a time lag which explains the existence of a large pool of detritus in most ecological systems. As already mentioned, the decomposition of this material is principally due to bacterial (and in terrestrial environments also fungal) activity. In the following sections we will explore some aspects of this key role of microorganisms.

3.1. DISSOLVED ORGANICS IN AQUATIC SYSTEMS

The distinction between dissolved and particulate organic matter is arbitrary since the whole size spectrum from small molecules (e.g., glucose) to very large macroscopic particles may be found in the sea. Usually dissolved organics are defined operationally as material which will pass filters with pore sizes of $0 \cdot 2$ or $0 \cdot 45 \mu m$. Bacteria exclusively take up small molecules; high molecular weight substrates are first hydrolyzed by extracellular (or membrane bound) enzymes before being utilized. From a physiological point of view the above definition of dissolved organics is, therefore, not very meaningful. However, the composition and microbial turnover is easier to study and consequently better known in the case of dissolved organics in aquatic systems, so we will discuss it in some detail in the following.

3.1.1. The Origin, Composition and Amount of Dissolved Organics

Since the beginning of the century it has been recognized that in the oceans and fresh water systems, the greatest part of the total organic material is present in the dissolved form. It is generally estimated that in the oceans, dissolved organics, particulate organics and living biomass occur in the proportions 100:10:2 (Parsons, 1963). In seawater, the concentration is usually of the order $1 \, \mathrm{mg} \, l^{-1}$, in inshore waters values may be 10 times higher and in lakes values between 1 and $10 \, \mathrm{mg} \, l^{-1}$ are usually reported with extreme values of up to $50 \, \mathrm{mg} \, l^{-1}$ (Jørgensen, 1966, 1976; Fenchel and Jørgensen, 1977; Ogura, 1975; Wetzel et al., 1972). For the oceans this means that the total pool of dissolved organics constitutes about $1 \cdot 5 \times 10^{18} \, g$. The annual primary production in the world oceans is believed to be of the order of $5 \times 10^{16} \, g$ organic matter. Not all of this production will appear as dissolved organics but as discussed below, 10% is a very conservative estimate. The turnover time of the dissolved matter is, therefore, somewhere within the range of 30–300 years. However, this is only an average value for a heterogeneous assemblage of organic molecules, some of which have an extremely rapid turnover (e.g., glucose) and others which are very resistant to microbial degradation (e.g., humic substances) and which have turnover times that may be measured in millenia. These considerations also do not take the physical heterogeneity of the oceans into account.

The quantification of the different constituents of dissolved organic matter is in general difficult since concentrations of the chemically well-defined and ecologically important low molecular weight compounds are very small. Values reported in the literature are not independent of the analytical methods employed. The reported ranges of some groups of compounds are shown in Fig. 14. The highest values derive from estuaries and interstitial water of sediment, the vertical bars indicate more typical offshore values. It can be seen that amino acids, soluble sugars and fatty acids constitute only a few percent of the total pool of dissolved organics. The dominating part consists of high molecular weight components and colloidal matter including carotenoids, waxes, humic material, and bound amino acids.

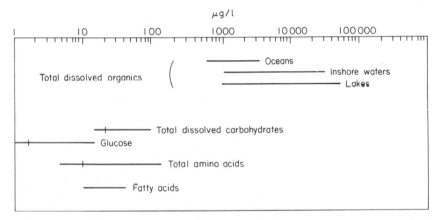

Fig. 14. The range of concentrations of total dissolved organics and of some low molecular weight components in natural waters. Vertical bars indicate typical values for seawater.

The dissolved organics derive from several sources. The most important single source may be the excretion of dissolved organics from algae during photosynthesis. Thus from 10 to 30% of the assimilated carbon in marine phytoplankton is lost as dissolved organics in the form of sugars, amino acids and especially glycollate. With respect to aquatic macrophytes it has been found that 4–6% of the production is lost as dissolved organic carbon. Other sources of dissolved organics are the autolysis of dead algal and animal tissue, excretion from animals and bacteria and terrestrial run-off (Johannes and Satomi, 1967; Riley, 1970; Hargrave, 1971; Otsuki and Hanya, 1972; Wetzel *et al.*, 1972). Thus a very substantial part of the production in aquatic environments is channeled through the pool of dissolved organic matter. We do not know exactly how large a fraction this is; it is certainly more than 10% and may well exceed 30% of the total biological production of the sea.

3.1.2. The Kinetics of Uptake of Dissolved Organics

Before discussing the uptake of dissolved organic material in natural waters it will be useful to give a brief discussion on the formal description of the uptake kinetics of dissolved carbon sources. This description is the basis for the experimental study of the uptake in natural environments.

Bacteria (and other cells) have active uptake mechanisms for certain organic and inorganic substrates. These mechanisms may be saturated so that with increasing substrate concentration the relative increase in uptake rate decreases until a maximum uptake velocity is reached. It has been shown empirically that under a given set of constant environmental conditions the uptake of a given organic substrate by a bacterium may be described by first order enzyme kinetics (Michaelis–Menten kinetics) so that

$$V = V_m[S/(K_m + S)],$$

where V is the uptake velocity, V_m the uptake velocity at saturation, S the substrate concentration, and K_m a constant measuring the value of S for $V = V_m/2$. Thus the uptake kinetics are completely described by the constants V_m and K_m. A high value of V_m signifies a high uptake capacity and a low value of K_m signifies a high substrate affinity; i.e., the uptake rate approaches V_m at relatively low substrate concentrations. The relation between S and V can be linearized by plotting the reciprocal functions (a Lineweaver–Burk plot, see Fig. 15). From the data points, values of V_m and K_m can therefore be found.

In nature the values of V_m and K_m for different bacteria are correlated for a given substrate so that strains with high V_m values also have high K_m values and conversely strains with low values of V_m also have low values of K_m. The ecological significance of this is that bacteria may either specialize in having a high uptake rate at high substrate concentrations (high values of the kinetic constants) or in being efficient competitors at low substrate concentrations (low values of the kinetic constants). Figure 15 shows two hypothetical strains with different values of kinetic constants. It can be seen that strain 2 is more efficient at low substrate concentrations whereas strain 1 is more efficient at high substrate concentrations.

Since there is a close relation between substrate uptake and population growth a similar relation between the specific growth rate, μ, the maximum growth rate, μ_m, the half saturation constant, K_m and the substrate concentra-

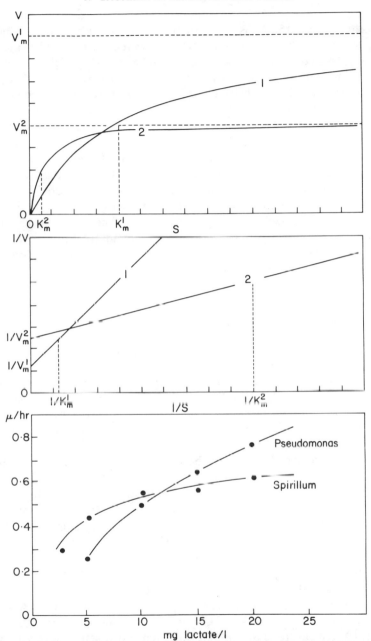

Fig. 15. Above: the uptake of a substrate as function of concentration, S, by two hypothetical strains of bacteria. Middle: a Lineweaver-Burk plot of the curves shown above. Below: the growth rates of two bacterial strains in a chemostat as a function of lactate concentration (redrawn from Veldkamp and Jannach, 1971).

tion S is found (Fig. 15). Thus the values of the constants can tell us which of the two strains will be competitively superior, given a certain substrate concentration. This concept of slow growing bacteria, with a high competitive ability at low substrate concentration, compared to bacteria with a high potential growth at high substrate concentrations, is related to the concept of the *autochthonous* versus the *zymogeneous* microflora of the soil bacteriologists. For a thorough discussion of these concepts see Veldkamp and Jannasch (1972).

The introduction of these concepts also allows a further analysis of the dynamics of dissolved organics in natural waters to be made, although a number of reservations must be taken. The above description of bacterial uptake kinetics does, strictly speaking, only apply to bacteria growing in a constant and homogeneous environment such as a chemostat. As discussed in more detail below, it has to be kept in mind that nature is not a chemostat and application of chemostat theory cannot uncritically be employed for natural conditions (Jannasch, 1974).

3.1.3. The Turnover and Mineralization of Dissolved Organics

In Section 3.1.1 it was indicated that dissolved organic matter consists mostly of a pool of macro-molecular and colloidal material which is very resistant to microbial degradation and a much smaller pool of low molecular weight compounds which can be directly used by bacteria. When freshly sampled seawater is taken into the laboratory, a relatively rapid initial rate of mineralization of smaller molecules is followed by a much lower rate of mineralization of resistant compounds (Ogura, 1975).

The uptake rate of radioactively-labeled compounds in samples of natural waters has shown that a substantial part of the carbon flow of aquatic systems passes through dissolved low molecular weight organics. Thus, these compounds which, due to bacterial activity, typically occur in concentrations of the order of $1-5\mu g\,l^{-1}$, have a very rapid turnover. The techniques used in these studies and their results are described in Wright and Hobbie (1965), Hobbie (1967, 1971), Vaccaro *et al.* (1968), Hobbie and Crawford (1969), Wood (1973), Crawford *et al.* (1974), and Wright and Shah (1975).

The principle of the method is to incubate water samples with a ^{14}C (or ^{3}H) labeled substrate (usually glucose, amino acids and fatty acids). After an incubation period, ^{14}C-labeled CO_2 and particulate (bacterial) ^{14}C are quantified as a measure of the uptake. Since natural substrate levels are very low and since carrier-free ^{14}C-labeled organic compounds cannot be obtained, most studies have not, so far, been real tracer experiments; i.e., the labeled material increases the chemical concentration and thus the bacterial uptake rate significantly relative to the natural levels. (Most recently, work using ^{3}H-labeled compounds at tracer level has been carried out.)

The solution to this problem has been attempted by incubating a series of subsamples of the water with an increasing amount of the same substrate; this will yield an estimate of V_m. It is assumed that the total microbial uptake in the water sample follows the Michaelis–Menten kinetics. By observing (see Section 3.1.2) that

$$(S_n + S_a)/V = (K_m + S_n + S_a)/V_m$$

where S_n and S_a are the natural and the added substrate concentrations respectively, it can be seen that plotting the ratio: (added radioactivity/assimilated radioactivity) against S_a will yield a straight line. Extrapolating the line to $S_a = 0$ will then yield the turnover time, T, at the natural substrate concentration, S_n (Fig. 16). If S_n is known, then the true uptake rate can be calculated.

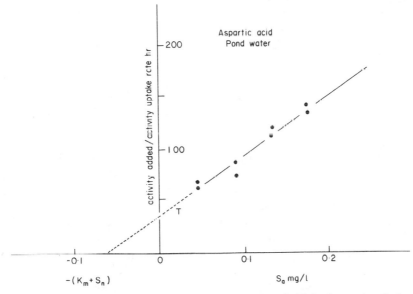

Fig. 16. The turnover time of an organic substrate as function of added substrate in a freshwater sample (redrawn from Hobbie, 1971).

The method has a number of shortcomings which have been discussed by Wright (1973), Williams (1973), and Baross *et al.* (1975). The most important ones are the following:

(1) It may be very difficult to quantify the natural substrate concentrations. In several studies this has not even been attempted but the turnover time and

V_m are given as measures of microbial activity and "heterotrophic potential" respectively.

(2)   The total number of important compounds and their relative importance is incompletely known. In actual practice a few "key compounds" such as common amino acids, glucose (being the hydrolysis product of cellulose and starch), acetate and glycollate have mainly been studied. It will probably be impossible to test all potentially important substrates or the hydrolytic products of all naturally occurring macromolecules. Already, for this reason, the method cannot give a complete picture of the heterotrophic activity of bacteria.

(3)   The assumption that the uptake of a given substrate by the entire microbial community will follow the Michaelis–Menten kinetics is in fact not warranted. If the different types of bacteria present have different uptake constants, then the measured uptake cannot be described by this relationship. It is to be expected that bacteria in a given water sample will have different kinetic constants since, even in a pelagic biota, nutrient concentrations may have a heterogeneous distribution due to adsorption to particles and different strains may have adapted to this heterogeneity. Also, bacteria may be capable of taking up more than one type of substrate but have different affinities for each of them. A reciprocal graphical representation of the effect of saturation will most likely approach linearity in any case, but the extrapolation back to the true turnover time may be in considerable error (Williams, 1973).

In spite of these difficulties the method has given us the best picture so far of the dynamics of dissolved organics in natural waters. It has been shown that values of the turnover time, T, and V_m vary considerably between different water bodies and seasonally within the same system and peaks in uptake often follow peaks in primary productivity. Thus, turnover times ranging from as little as 0·5 to more than 10^4 hr been found when comparing aquatic habitats ranging from polluted ponds and productive estuaries to oligotrophic arctic lakes. In a Swedish lake the turnover time for glucose varied from less than 10 hr in summer to more than 1000 hr in winter. Peaks of V_m correlate clearly with peaks in phytoplankton productivity. In marine habitats glucose uptake seems to vary from 0·002 $\mu g\,l^{-1}\,hr^{-1}$ in oceanic waters to 0·25 $\mu g\,l^{-1}\,hr^{-1}$ in coastal waters. In an estuary it has been found that the bacterial production, based on the uptake of amino acids alone, constituted about 10% of the phytoplankton production. This latter figure shows clearly the significance of dissolved organics and their mineralization through microbial activity. Glycollate is of special interest since it is a major excretion product of phytoplankton. Studies have shown that the majority of bacterial isolates are capable of utilizing this compound. Its turnover time in coastal waters has been found to vary from 7 hr in surface waters to 2000 hr in the deeper, aphotic zone.

The percentage of the substrates taken up which is respired varies according to the substrate. Thus in the case of glucose and leucine, values of 10–20% have been reported, whereas in the case of, e.g., aspartic acid and glycollate 50–70% of the material taken up is respired.

When natural waters are enriched with substrates such as glucose, the microflora quickly respond by increasing their heterotrophic potential, V_m. While this has not yet been studied in detail, the effect could be due to an increase in cell numbers, to an adaption of individual cells, or to qualitative changes in the composition of the microbial flora through competitive interactions.

It can, of course, be asked to what extent organisms other than bacteria are responsible for the measured uptake of dissolved organics. A great many unicellular eukaryotes (algae, protozoa) and aquatic invertebrates have been shown to possess active uptake mechanisms for e.g., glucose and amino acids. Some non-pigmented algae living in environments very rich in dissolved organics, such as in decaying seaweeds, are known to depend on dissolved organics. Many small protozoa can be grown axenically in the laboratory on dissolved substrates, albeit mostly on concentrations far exceeding that usually found in natural waters. The question of dissolved organics as a source of food for animals has been dealt with in detail by Jørgensen (1966, 1976). It is concluded, that with the exception of special environments, dissolved organics play a minor role as food for animals and that the role of animals in turning over dissolved organics in natural waters in general must be trivial. The half saturation constant of uptake for different substrates is in general much higher than that found in bacteria; the latter, therefore, probably reduce the concentration to levels at which the substrates cannot be used efficiently by higher organisms.

Several authors have measured the substrate uptake by different sized fractions of pelagic microbial communities. Hobbie and Wright (1965), Wright and Hobbie (1966), Munro and Brock (1968), and Williams (1970) all found that the fraction $< 3 \mu m$ (i.e., bacteria) is resonsible for the greatest part of the substrate uptake; only Allen (1971) found that the largest part was taken up by microflagellates. Hobbie and Wright (op. cit.) also found that K_m value of algae are considerably higher than those of bacteria. Finally correlations between uptake of organic solutes and bacterial numbers (in particular as plate counts based on media with the tested substrate) have been demonstrated (Gordon et al., 1973; Hamilton and Preslan, 1970; Hobbie et al., 1972). It is, therefore, warranted to conclude that in general bacteria are quite dominating with respect to the turnover of dissolved organics in pelagic environments.

It has already been mentioned that the uptake of some selected compounds will give an incomplete picture of the total heterotrophic activity of bacteria since there are a great number of potential substrates. This problem applies especially to aerobic conditions where the complete mineralization may take place within a single cell. Under anaerobic conditions, decomposition begins as under aerobic conditions by the hydrolysis of macro-molecules to low molecular weight compounds. These are then fermented to a relatively restricted number of compounds (lactic, formic, acetic, propionic, and butyric acids, H_2) which serve as substrates for other bacteria which mineralize them further by anaerobic respiration or methanogenesis. The turnover of a few key compounds may, therefore, give a much more complete picture of heterotrophic bacterial activity under anaerobic conditions.

For measuring the total heterotrophic activity under aerobic conditions a number of other methods have therefore been considered. One of these is the CO_2-dark uptake of water samples. Heterotrophic bacteria assimilate a certain amount of CO_2 for their cell carbon (Section 1.2.1). Assuming that this uptake is a constant fraction of cell growth (estimated to be about $6 \cdot 6 \%$), this uptake should be a measure of bacterial production (Sorokin, 1965; Kuznetzov, 1968). The method can obviously not be used when chemo-autotrophic or methane oxidizing bacteria are present since they obtain all, or a considerable portion of their cell carbon from assimilated CO_2. Also, heterotrophic bacteria in general are quite variable with respect to their CO_2 uptake (Overbeck, 1974). Another method is to use the assimilatory SO_4^{2-} uptake as a measure of bacterial growth. At least in seawater this essential ion is never limiting and the uptake should thus directly reflect bacterial production (Monheimer, 1974). These methods have not yet yielded much reliable information on the heterotrophic activity of natural waters.

Another important problem is the turnover of dissolved organic matter in the interstitial water of sediments. Here substrate concentrations and bacterial population sizes are much higher than found in the overlying free water and much higher uptake rates could thus be expected. Measurements of ^{14}C-glucose and acetate uptake have indeed indicated that this is so in the case of a lake sediment (Hall et al., 1972). However, it is very questionable whether this method will yield results which can be interpreted quantitatively when applied to sediments. It would seem most unlikely that, in a heterogeneous environment like sediments, the uptake will conform to the Michaelis–Menten kinetics although published results indicate this to be so. Secondly, it is necessary to stir the sediment, prior to or even during the incubation, in order to obtain a homogeneous distribution of the labeled substrates unless tracer levels can be used. This will almost certainly increase the microbial activity considerably, relative to the undisturbed sediment.

3.1.4. The Role of Solid Surfaces

Zobell (1943) showed that when seawater is placed in bottles, its oxygen consumption increases for some time and that this effect can be stimulated by the addition of inert particles (e.g., acid rinsed sand). These findings were interpreted to show that dissolved organic material is adsorbed to surfaces where bacteria can efficiently utilize the local increase in substrate concentrations. The experiment does indicate that the substrate concentration in seawater is generally limiting and that it is kept at this low level by bacterial uptake.

Jannach and Pritchard (1972) studied this effect in more detail using pure cultures and various types of inert particles. They could show that in dilute substrate concentrations, most bacteria are attached to the particles and the mineralization is more efficient in the presence than in the absence of particles. At high substrate concentrations, the addition of particles has no effect on the rate of mineralization and the bacteria are not attached to the particles. Bacteria are not found attached to particles in the absence of a substrate.

The significance of inert particles for bacterial activity in natural waters is not clear. Bacterial aggregates apparently based on clay minerals, diatom frustules, etc., have been described from the sea as well as from lake water (Paerl, 1974) and the stimulating effect of suspended silt for the bacterial activity in an oligotrophic lake has been claimed (Paerl and Goldman, 1972). Other workers, e.g., Hobbie et al., 1972) found that only a small part of the pelagic bacteria are associated with suspended particles.

Bacterial aggregates may be important for the carbon flow of aquatic systems in one other way. Many filter feeding animals cannot retain particles of bacterial size but may be able to retain and utilize bacterial aggregates and bacteria attached to suspended particles (Seki, 1972; Fenchel et al., 1975).

3.2. THE DECOMPOSITION OF PARTICULATE DETRITUS

Considerable amounts of detritus enter food chains in a particulate rather than in a dissolved form; the bulk of this material is plant litter and dead phytoplankton cells mixed with debris from dead animals and microbial cells. By far the greatest part of this material is eventually mineralized. An overall quantification of the rate of mineralization and turnover can, therefore, in principle be derived from a knowledge of the input of detritus (allochtonous material and locally produced plant tissue which is not consumed by herbivores) and the steady state amount of detritus. A more detailed picture is difficult to achieve due to the heterogeneous nature of particulate detritus.

This heterogeneity is evident with respect to the mechanical and chemical qualities of detrital particles, as well as the chemical environment and the associated decomposer organisms. In the following sections we will discuss the role of bacteria for the primary decomposition of particulate detritus and factors which determine the rate of microbial degradation.

It will be practical to discuss detrital decomposition in aqauatic environments separately from terrestrial environments. There are significant chemical and biological differences between the decomposer food chains in these two kinds of habitats. In addition, the types of problems studied by soil biologists and aquatic ecologists are in some respects different.

3.2.1. Particulate Detritus in Aquatic Environments

In most fresh water and estuarine environments the greatest input of particulate matter derives from macrophyte tissue (such as dead leaves or thallus of seagrasses, kelps, mangroves and terrestrial plants). Further offshore, dead phytoplankton cells play an increasing role.

The quantification of detritus in aquatic environments is extremely difficult since it undergoes cycles of deposition and resuspension and in shallow water areas its distribution may be very patchy and it is often sorted more or less according to particle size. Furthermore, particles in different stages of decomposition are often mixed. A large part of the material becomes buried in sediments and therefore undergoes a more or less complete mineralization under anaerobic conditions.

The great importance of detritus for the food chains in the sea was already established early in this century (e.g., Boysen–Jensen, 1914; Petersen, 1918). However, the role of microorganisms in detritus food chains was not appreciated. Many aquatic invertebrates, the "detritus feeders", ingest the detrital material of sediments or filter suspended detrital particles. It is now believed that these animals are only capable of assimilating little of the compounds constituting the bulk of the material (structural carbohydrates). The important food item for detritus feeders is constituted by the microorganisms associated with the detrital particles; bacteria, fungi and the bactiverous protozoa and micro-metazoans (Odum and de la Cruz, 1967; Fenchel, 1970, 1972, 1977; Fenchel and Harrison, 1976; Fenchel and Jørgensen, 1977; Mann, 1972, 1976, and references therein).

Bacteria are probably the most important primary decomposers in the sea but direct microscopical observations indicate that phycomycetes may also play a role and, e.g., Fell and Master (1973) and Meyers and Hopper (1973) have reported on various types of fungi involved in the decomposition of mangrove leaves, seagrasses and cellulose in marine habitats. In fresh water environments fungi are probably more important than in the sea, but to our knowledge, there does not exist any estimate of the quantitative role of fungi relative to that of bacteria as decomposers in aquatic environments.

Bacteria on the surface of particles of macrophyte origin are mainly responsible for the process of decomposition. Mechanical degradation of detritus through wave action, or by detritus feeding animals therefore stimulates decomposition by increasing the exposed surface area of the particles (Fenchel, 1970; Hargrave, 1970, 1976). On the surface of detrital particles the density of bacteria is mainly controlled by grazing of protozoa and other faunal elements. If protozoa are excluded from the microbial communities under experimental conditions, bacterial density increases by a factor of 2 to 10 but the rate of mineralization decreases relative to similar systems where the bacteria are grazed. The reason for this stimulation of bacterial activity by grazing protozoa is not quite clear; it is discussed in Fenchel (1977) and Fenchel and Jørgensen (1977).

Very little is known about the rates at which different chemical constituents of plant material are decomposed in the sea. Fresh leaves and thallus fragments initially leach soluble constituents which are rapidly mineralized by bacteria (Fenchel, 1972; Oláh, 1972) Fenchel and Harrison (1976) and Fenchel (1977) found that detritus (water extracted, homogeneously ^{14}C-labeled barley hay in seawater, with mineral nutrients and inoculated with a mixed microflora) is rapidly mineralized by what can be approximated by a first order process until at least 85 % of the material is mineralized (Fig. 20). This indicates that structural carbohydrates may be rapidly mineralized in seawater under aerobic conditions. Under similar conditions the decompostion of fresh seagrass leaves is much slower (Fenchel and Harrison, 1976). This may be due to the presence of substances inhibitory to bacteria (e.g., phenols) but this has not been studied.

Cellulose is formed by some marine plants and algae but this compound is much less important in the sea than it is in terrestrial environments. Among important structural polymers of marine algae, alginate (polymers of mannuronic and glucuronic acids) may be mentioned. Alginate is decomposed by several bacteria (*Alginobacter, Alginomonas, Nocardia,* and others). Polymers of sulfate esters of galactan and other sugars (carrageenan, iridophycan, furcellaran, agaragar) are especially produced by rhodophytes; they are hydrolyzed by bacteria belonging to genera of, e.g., *Cytophaga, Agarbacterium,* and *Nocardia.* Chitin is a structural polymer of fungi and of various animals, in particular arthropods, and is therefore common in aquatic as well as in terrestrial environments. It is a polymer consisting of β- 1, 4-linked, N-acetylglucosamine units; it is hydrolyzed by many bacteria including species of *Serratia, Pseudomonas, Vibrio, Flavobacterium,* and *Clostridium.* Proteolytic bacteria are very frequent in the aquatic (and terrestrial) environment. Proteins are hydrolyzed by exo-enzymes to oligopeptides and taken up by the cells. The peptides are further hydrolyzed to amino acids and are used for growth or are deaminated to release ammonia. Fats are also utilized by many bacteria possessing lipases. Hydrolysis produces glycerol and fatty acids; the latter are utilized by splitting them to acetate units. Various waxes and oils are utilized in a similar way. The degradation of hydrocarbons will be discussed in Section 3.5. (For references, see ZoBell and Upham, 1944; Campbell and Williams, 1951; Skerman, 1967; Doetsch and Cook, 1973; and Rheinheimer, 1974).

Litter bag experiments in aquatic habitats have shown a wide variation in decomposition rates according to plant species. Thus the time taken for the disappearance of 50% of the material varied from 14 days to more than 6 months for different species (Wood *et al.*, 1969). Saunders (1976) who gives a general review of decomposition in freshwaters, also compiles data on the decomposition of various limnic and terrestrial plant species.

3.2.2. The Decomposition of Soil Organic Matter

The mineralization of soil organic matter has been studied in much more detail than mineralization of detritus in aquatic systems. This, of course, is in part due to the fact that the subject is of paramount importance in agricultural science and in forestry. Thus much more is known about the taxonomy of the microorganisms responsible for mineralization in soils and on the decomposition rates of different constituents of litter. We cannot review the large amount of literature on the subject here; the reader is referred to Burges and Raw (1967), Gray and Williams (1971), Dickinson and Pugh (1974) and Richards (1976). Only an outline of the most important characteristics of decomposition in soils will be given here.

The chemical composition of terrestrial plant litter differs from that of marine detritus, the former containing more cellulose and xylan and especially compounds like lignin, suberin, cutin, phenols and resins. In addition to plant litter, root exudates add to soil organic matter. The environmental conditions also differ from those of aquatic habitats; most importantly, humidity plays a large role in the decomposition of soil organic matter. Whereas in aquatic environments fungi are believed to be of secondary importance they are of considerable significance in terrestrial habitats.

The decomposition of leaves and twigs is initiated already before they enter the soil by a special—mainly fungal—flora associated with living or senescent plant surfaces. In the soil, leaching of low molecular weight compounds takes place; this material is fairly rapidly mineralized by bacteria and "sugar fungi", i.e., mainly yeasts. The further succession of the microflora also depends on environmental conditions; in acid soils fungi are initially most important, followed by bacteria during the later stages of decomposition. In more alkaline soils bacteria initially play a larger role. Fungi—in particular basidiomycetes—are generally believed to be the most important decomposers of structural plant compounds, e.g., cellulose and especially lignin. The invasion by fungi is then followed by bacteria which mineralize fungal exudates and dead hyphae. However, cellulolytic bacteria, as mentioned below, are well known from soils.

Cellulose may be the most common organic compound on earth. It is a polymer of β-1, 4-linked D-glucose units and with molecular weights ranging between 1 and 2×10^6. Wood consists of 40–50% and straw of about 35% cellulose. Cellulose hydrolysis is widespread among bacteria; altogether 78 species, representing among others the genera *Cellulomonas*, *Vibrio*, *Pseudomonas*, *Streptomyces*, *Cytophaga* and *Sorangium* (aerobically) and *Clostridium* (anaerobically) can utilize cellulose. In many forms, the cellulases are membrane bound; in other cases extracellular celluloses can be detected in culture media. Two main types of cellulases are found, i.e., those which hydrolyze the glucose bonds at random in the cellulose molecule and those hydrolyzing terminal bonds producing glucose or cellobiose (a disaccharide). Finely ground cellulose (with a large total surface area) may be rapidly decomposed by bacteria in the presence of mineral nutrients; in nature where cellulose occurs as large particles (wood, dead leaves) and in a matrix with other structural compounds (lignin, xylan) the bacterial decomposition takes place mainly from the surfaces and cellulose degradation is a relatively slow process.

The term "hemicellulose" comprises a number of polymers of various sugars (xylan, which is a β-1, 4-linked xylose polymer, and also various galactanes, arabanes and heteropolymers of glucuronic and galacturonic acids). Wood contains about 35% hemicellulose; the corresponding figure for straw is about 20%. Pectin is a polymer of galacturonic acid with some degree of esterification; it is widely distributed among vascular plants. All these compounds are readily hydrolyzed by many bacteria, also under anaerobic conditions, as are the storage carbohydrates of vascular plants: amylose and amylopectin which are unbranched and branched polymers respectively of glucose.

Animals, fungi and bacteria themselves also contribute polysaccharides and related polymers to the pool of particulate detritus including glycogen, chitin, hyaluronic acid, poly β-hydroxy butyrate, and uronic acid polymers; these are broken down by various microorganisms. (The degradation of carbohydrates in connection with symbiosis is discussed separately in Section 8.1.)

Lignin makes up about 20% of wood and somewhat less in other tissue of vascular plants. Lignins are amorphous polymers of aromatic (e.g., guaicyl-glycerol and phenyl propane) subunits. The structure of lignin is not known in detail and seems to differ for different plants. Also, pure preparations of lignin from a given plant tissue differ somewhat according to the extraction method employed and the decomposition of lignin is, therfore, difficult to study (Pearl, 1967). Bacteria (*Psuedomonas* sp.) which degrade lignin have been isolated (Sørensen, 1962) but in nature this process is believed to be carried

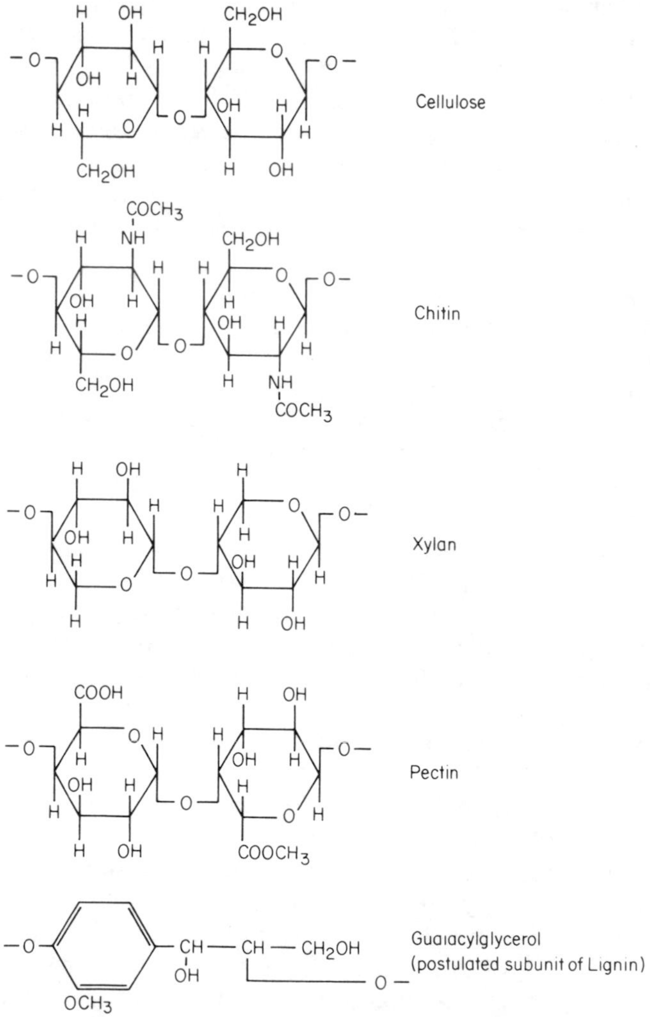

Fig. 17. Some structural carbohydrates and a possible subunit of lignin.

out mainly by fungi. Lignin degradation, at least the cleavage of the aromatic rings, seems to require oxygen (but see Section 3.5) and, therefore, does not take place under anaerobic conditions. Lignin degradation is always a slow process. Phenols are produced by a number of plants and are degraded by species of *Nocardia* and *Pseudomonas*. The function of phenols as well as of a very great number of other secondary plant compounds is probably to limit herbivory and microbial attack. Yet, they are probably all degraded by at least one type of microorganism although mineralization may in some cases take place very slowly. The chemical composition of some structural plant polymers is shown in Fig. 17. For references on the degradation of soil organics see Alexander (1961), Tribe (1961), McLaren and Peterson (1967), Gray and Parkinson (1968), and Doetsch and Cook (1973). The mineralization of hydrocarbons is discussed in Section 3.5.

Figure 18 shows, in a somewhat idealized way, the decomposition rates of some of the most important litter components in a soil. In fact, the decomposition rates vary considerably according to the type of terrestrial ecosystem. Thus in tropical forests, mineralization rates of up to 40% per month of the detritus pool present have been recorded whereas in tundras less than 20% per year is mineralized (Witkamp and Ausmus, 1976).

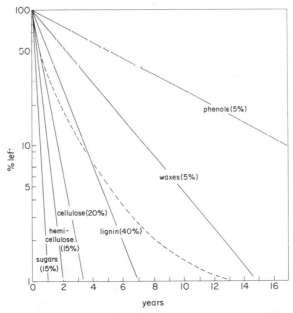

Fig. 18. An idealized presentation of the breakdown rates of some important litter components in soil. The broken line is a sum curve showing the decomposition of the total litter organics. The numbers in parentheses indicate the approximate quantitative importance of the components in litter (redrawn from Stout *et al.*, 1976).

3.3. MINERAL NUTRIENTS AND DECOMPOSITION

An important aspect of decomposition is that mineral nutrients (e.g., N, P, K) bound in dead organic matter are released in an inorganic form, available for the primary producers. However, the relationship between decomposition and mineral nutrients may be more complex. It has long been known that when nutrient-poor organic substrates (such as pure cellulose or straw) are added to soils the immediate effect may be an immobilization of mineral nutrients. Bacteria have a relatively high content of essential elements relative to plant tissue. Bacteria (and fungi) therefore assimilate dissolved inorganic nutrients simultaneously with the uptake of organic substrates. Fresh leaves may have C/N ratios as high as 100 and fresh wood even up to 1000 (Witkamp and Ausmus, 1976). As examples of C/N ratios of aquatic plants, large brown algae have values of 16–68, seagrass 17–70, and rhodophyceans about 20. Microalgae have lower values, e.g., 6–7 for diatoms and chlorophyceans. Bacteria characteristically have values around 5. With respect to phosphorus, typical values of the C/P ratio are, for aquatic macrophytes around 200 whereas bacteria have values around 20.

Bacteria utilize their organic substrates for the synthesis of new cell material and for their energy matabolism. Assume that a bacterium has a C/N ratio C_b/N_b where C_b and N_b are the amounts of carbon and nitrogen in one bacterial cell. In order to produce biomass with the carbon content C_b (viz., to produce one new cell) the bacterium has to take up $C_a = C_b + C_r$ organic carbon, where C_r is the respired carbon so that C_b/C_a is the growth yield. In order for the substrate to satisfy the need for nitrogen its C/N ratio should therefore not exceed

$$(C_b + C_r)/N_b = (C_b/N_b)/(C_b/C_a),$$

i.e., the C/N ratio of the bacterial cytoplasm divided by the growth yield. If the substrate has a higher C/N value a net immobilization of dissolved nitrogen will take place; conversely if the substrate has a lower C/N value, a net mineralization of nitrogen will take place (Fig. 19).

Growth yields of aerobic bacteria vary according to the substrate, growth rate and other factors; in the literature, values between 20 and 80% have been reported for different strains growing on different kinds of substrates. If we accept 50% as a reasonable estimate of average growth efficiency and a C/N ratio for bacteria of 5, the net mineralization should only take place when the C/N value is lower than about 10.

Agronomic experience, however, indicates that net mineralization will take place already when the substrate C/N ratio is lower than about 20 (Richards, 1976), suggesting a somewhat lower growth yield ($\sim 25\%$) for the average soil microorganism or that the portion broken down has a higher value of C/N. In anaerobic metabolism the growth yield is considerably lower than in aerobic metabolism, i.e., within the range 5–30% (see Section 1.1). In anoxic environments, therefore, the availability of mineral nutrients relative to the availability of carbon substrates is higher.

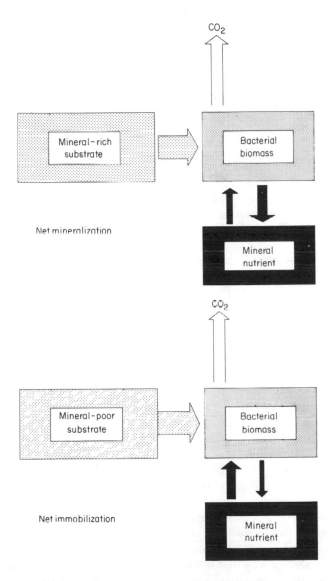

Fig. 19. A schematical presentation of mineral uptake and release of mineral nutrients by bacteria decomposing mineral-rich and mineral-poor substrates respectively.

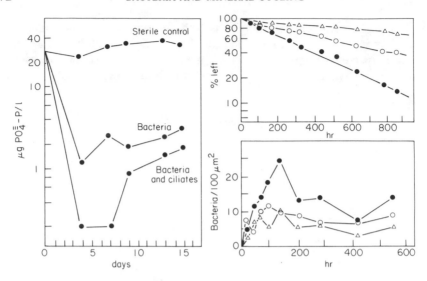

Fig. 20. Left: the change in PO_4^{3-} concentration in a microcosm with decomposing barley hay in seawater and inoculated with bacteria, bacteria and bacterivorous ciliates, and a sterile control respectively. Right, above: the decomposition rate of ^{14}C-labeled hay in microcosms with seawater enriched with PO_4^{3-} (△), NO_3^- (○), and with $PO_4^{3-} + NO_3^-$ (●), respectively and with a mixed microbial inoculate. Below: the bacterial population sizes (given as number of cells per unit surface of the detrital particles) in the systems shown above. The effect of enrichment was also evident from the population sizes of the consumers of the bacteria (zooflagellates and ciliates) in the systems. (All figures redrawn from Fenchel and Harrison, 1976.)

Figure 20 shows an example of immobilization of PO_4^{3-} during the decomposition of barley hay in seawater. It is seen that the available phosphate is rapidly depleted to very low levels. After about a week, when a substantial part of the organic substrate is decomposed, a net mineralization of inorganic P is detectable. Barsdate *et al.* (1974) found, using tracer experiments with $^{32}PO_4^{3-}$, that in such systems there is a rapid uptake and excretion of inorganic P by bacteria, a rate which far exceeds the actual mineralization rate of the substrate. Studying the decomposition of *Carex* litter in flasks with freshwater and inoculated with a mixed Microflora, they found that $0·84\,\mu g$ $P\,l^{-1}\,hr^{-1}$ were being mineralized from the litter after about a week. Simultaneously, however, the bacteria took up (and excreted) about $150\,\mu g\,P\,l^{-1}$ hr^{-1} in the system.

In microcosms such as the ones described above, enrichment with inorganic nutrients up to levels much higher than usually found in nature, will increase the decomposition rate and cell biomass (Fig. 20). These experiments indicate in a general way that the availability of nutrients may limit the decomposition rate in nature. However, they cannot directly be used for quantitative predictions on the effect in natural ecosystems. Thus, it will depend

on the growth yield, the rate of growth and the actual microbial biomass involved in breaking down a given substrate· It can, in general, be predicted
that the substrates broken down more easily will result in a higher degree of
mineral immobilization than will a more slowly decomposing substrate such
as lignified tissue.

The immobilization of mineral nutrients by decomposer microorganisms
has several important ecological implications. Plant-derived, detrital particles often have a content of mineral nutrients which is too low to serve as
food for animals. The microbial growth on the surface of such particles will
enrich them with mineral nutrients and thus increase the nutritive value for
detritus feeders and browsing animals (for aquatic systems see Odum and
de la Cruz, 1967; Hynes and Kaushik, 1968; Mann, 1972; Fenchel and
Harrison, 1976; Fenchel and Jørgensen, 1977 and references therein; see
also Fig. 21).

Another important aspect is that the immobilization of nutrients will tend
to retain them in the soils outside the growing season of plants, whereas
inorganic nutrients, especially NO_3^-, otherwise tend to be washed out of the
soils (e.g., Witkamp and Ausmus, 1976).

Finally it is an important question whether, or to what extent, the availability of mineral nutrients limits the rate of mineralization in natural eco-

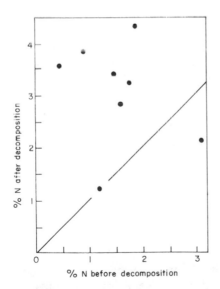

Fig. 21. The percentage of nitrogen in particulate, partly decomposed detritus derived from
marine macrophytes compared to the nitrogen contents of similar, undecomposed plant tissue.
Points over the line indicate an enrichment with nitrogen during decomposition (redrawn from
Fenchel and Harrison, 1976).

systems. There are examples to show that the addition of, e.g., inorganic N to natural waters increases the mineralization rate of cellulose or leaf litter but the quantitative importance is not completely understood in any kind of ecosystem. It is possible that, just as the availability of N and P in many systems limits the primary production, these elements also control the mineralization rate. Thus, immobilization and release of nutrients may in part control the relation between decomposition and production. The large steady-state pool of dead organic matter, characterizing many ecosystems, may then reflect the fact that plant tissues have lower nutrient contents than do the decomposer organisms and that the primary producers and decomposers compete for essential nutrients. The large pool of dead organic matter may, however, reflect the low rate at which certain structural plant polymers can be hydrolyzed even under optimal conditions and thus explain the fact that decomposition lags behind production.

3.4. HUMIFICATION AND FOSSILIZATION OF ORGANIC MATTER

As organic matter is mineralized, the portion remaining becomes increasingly resistant to microbial attack. Also, this material becomes chemically altered due to microbial and abiological processes. These remains, usually termed humic substances, have an increasing turnover time as a function of age and, therefore, tend to accumulate, comprising a large fraction of the organic material of the soils, sediments and water. Under certain environmental conditions, some of these materials will never mineralize but will be incorporated into sediments. They are then further metamorphosed through abiotic processes to become the organic components of sedimentary rocks (kerogen) and fossil fuels (petroleum, lignite, coal). We will give only a very short and superficial treatment of these processes here; in particular of the diagenetic processes leading from humic materials to fossilized organics belong to geochemical sciences more than to microbiology. We will give a short review of some microbial aspects and otherwise refer the reader to Eglinton and Murphy (1969), Swain (1970), Gjessing (1976), textbooks in soil science (e.g., Gray and Williams, 1971) and references therein (see also Section 4.4). Another aspect of chemical fossilization is the accumulation of reduced and elemental sulfur; this will be discussed in Section 6.2.1.

In terrestrial soils, the process of humification has been well studied. In temperate forest soils, the initial transformations can be followed in the vertical zonation with discrete layers corresponding to the annual litter falls. This transformation is first of all characterized by a decrease in the more easily decomposable constituents (e.g., carbohydrates, including cellulose, and proteins) and a relative increase in the more resistant constituents (e.g., lignin, cork substances, resins). This material is then, through microbial as well as abiological processes, transformed into humic substances. These are normally recovered from soils by alkali extraction. The acid-insoluble fraction of this extract is termed humic acid and the acid-soluble fraction is termed fulvic acid. The alkali-insoluble fraction of soil organic matter is termed humin. There are no sharp distinctions between these fractions, the differences mainly being the molecular weight and the attached side groups; humin is believed to consist of humic and fulvic acids bound to mineral matter.

The core of humic substances is made up of aromatic rings; these derive from lignin residues, phenols and quinones synthesized by microorganisms and later polymerized together with nitrogenous compounds to form the humic substances. Experiments with ^{14}C-labeled substrates (microbial carbohydrates, cellulose, glucose and wheat straw) added to soils, show that a part of the labeled carbon will rapidly turn up in the humic fraction (Martin et al., 1974). This probably represents microbially produced amino acids which are bound to the humic fraction. Sørensen (1975) showed that when ^{14}C-labeled cellulose is allowed to decompose in soil, some of the bacterially synthesized amino acids become bound to clay minerals and are stabilized in this way for many years.

The microbial degradation of humic substances is difficult to study since it is a very slow process, the precise chemical composition of the substrate is not known, and the decomposition may be restricted to various attached compounds (e.g., amino acids), whereas the aromatic core is left. It is believed that the degradation of humic substances in soils is largely due to basidiomycetes and ascomycetes.

The use of ^{14}C dating, and the use of the fact that additional ^{14}C was incorporated into plants following the hydrogen bomb tests in the 1950's and early 1960's, have given some information on the turnover time of humus in different soils. Turnover times exceeding 1000 years have been demonstrated in natural soils; cultivation of a soil may decrease this figure considerably. In general, the tendency to accumulate humic material in soils is higher in humid and cold climates, while mineralization is more efficient in tropical soils.

Under anaerobic conditions, in water-logged soils and swamps, the mineralization of resistant plant residues including waxes, resins, cork substances and lignin, is very inefficient resulting in low pH and peats, which often preserve the original structure of plant tissue. Aromatic rings, such as are found in lignin, seem to require oxygen for biologically catalyzed cleavage. Also basidiomycetes seem to be the most important primary decomposers of lignin and they are essentially aerobic organisms. Finally, in conjunction with the low pH values, peat seems to contain substances inhibitory to bacteria so that even animal tissue may be preserved. The nature of these substances is poorly known. In addition to humic substances, peat contains lignin, some cellulose and bitumen, which consists of waxes, paraffins and resins. Over geological time peat may through abiological processes turn into lignite (brown coal) and eventually into hard coal.

In marine environments there are two important sources of detritus; the remains of phytoplankton and algae on one hand and the remains of vascular plants and terrestrial material (including dissolved humic substances) on the other hand. The latter type of material will, under anaerobic conditions, form peat (e.g., in mangroves and in dense strands of seagrasses). Some of the constituents of this part of detritus seem also to be quite resistant under aerobic conditions and may make up a substantial part of the dissolved humic substance of seawater.

The components deriving from algae and phytoplankton are, under aerobic conditions, believed to mineralize nearly completely. Under anaerobic conditions these components may in part accumulate together with the more resistant components and derivatives from vascular plants, although a slow mineralization of sediment organics take place deep in the sediments over the centuries by methane producers and sulfate reducers. The hydrocarbons, constituting petroleum, derive especially from fatty acids (lipids, waxes) through abiotic processes which are still incompletely understood.

3.5.　THE MINERALIZATION OF HYDROCARBONS

The changes leading from peat to anthracite coal imply an increase in carbon content (from about 55 to 94 %) and this carbon can probably not return to the biosphere unless it is combusted directly to CO_2. Fossil fuels in the form of hydrocarbons (petroleum and natural gas), however, may be mineralized through the activity of microorganisms. This could at least in part be expected since many hydrocarbons (e.g., methane, terpenes, camphor, carotenoids, paraffins) are produced by living organisms.

Crude oils contain normal paraffins ranging in length from 1 to 30C, isoparaffins, branched paraffins (e.g., phytane, pristane), cycloalkanes, aromatic hydrocarbons, and steranes; these occur in various proportions together with some nonhydrocarbons, e.g., metalloporphyrins. Many microorganisms (species of the genera *Pseudomonas, Flavobacterium, Alcaligenes, Achromobacter, Nocardia, Mycobacterium, Arthrobacter, Micrococcus,* and *Brevibacterium* in addition to certain yeasts) are known to utilize hydrocarbons. Different bacteria show some degree of substrate specificity with respect to different types of hydrocarbons. In general, unbranched hydrocarbons are more easily degraded than branched ones; these are again more easily degraded than cyclic hydrocarbons, and aromatic compounds show the slowest rate of mineralization. Around natural and artificial seeps of gas and petroleum and tar pits, high concentrations of hydrocarbon-utilizing bacteria may be found. In connection with petroleum seeps, *ozokerite* may be formed; it is a waxy substance consisting of the higher alkane fractions mixed with bacterial products and from which *Mycobacterium* and *Nocardia* may be isolated (Doetsch and Cook, 1973). Some bacteria associated with surface water films have been found to utilize petroleum vapours (Hirsch and Engel, 1965). It has been found that certain oil degrading bacteria have hydrophobic cell surfaces (Heyer and Schwartz, 1970).

Oil degrading microorganisms have attracted much interest since they are the main agents for removing the effects of oil spills (see e.g., Fuhs, 1961; Zobell, 1964; Floodgate, 1972; Anhearn and Meyers, 1973; Jensen, 1975; Lehtomäki and Niemalä, 1975). Of paramount importance for rapid mineralization is dispersal, such as sorption to inert particles, in order to increase the surface area on which microbial degradation can take place and to ensure the availability of oxygen. The availability of mineral nutrients is also important. Various attempts to stimulate microbial degradation of oil spills have been made; these include fertilization with nutrients and seeding the spills with hydrocarbon-degrading microorganisms.

It now seems generally accepted that the bacterial degradation of non-cyclic hydrocarbons is, in principle, initiated by the oxidation of the terminal methyl group with molecular oxygen and catalyzed by an oxygenase. The corresponding alcohol is thus formed. This is then further oxidized to the corresponding aldehyde and then to the fatty acid which is finally converted into acetate units, which can be utilized by the cell in the normal manner. Some variations on this theme are known. Cyclic and aromatic hydrocarbons are also first oxidized with O_2, followed by ring scission. Thus, hydrocarbon degradation seems most probable under aerobic conditions. Facultative anaerobic nitrate reducing bacteria like *Pseudomonas* cannot utilize hydrocarbons in the absence of O_2 (Kallio *et al.*, 1963; van der Linden and Thijsse, 1965; van Ravenswaay *et al.*, 1971; Doetsch and Cook, 1973).

An anaerobic oxidation of certain short chain alkanes, however, has been suggested for *Pseudomonas* (Senez and Azoulay, 1963). In this case dehydrogenation of the terminal CH_3-group takes place by a NAD-dependent dehydrogenase, in the presence of the redox pigment pyocyanin, to yield the corresponding olefin. Oxygen is then introduced by a reaction with water to form an alcohol. A following dehydrogenation and reaction with water then leads to the formation of a fatty acid. The importance of this pathway is not known. Other authors have not been able to reproduce these results; for a critical discussion, see van der Linden and Thijsse (1965).

As in the case of methane, various geochemical and circumstantial evidence indicates that hydrocarbon oxidation may take place under anaerobic conditions through sulfate reduction. Short chain hydrocarbon utilization by sulfate reducers in the laboratory has also been reported by some authors. There is also evidence showing that some aromatic compounds may be degraded anaerobically through nitrate reduction or methanogenic pathways (Guyer and Hegeman, 1969; Nottingham and Hungate, 1969; Taylor and Heeb, 1972; Ferry and Wolfe, 1976).

CHAPTER 4

The Carbon Cycle

The C-, N-, and S-cycles share the common property of having an eight electron difference between the most oxidized and the most reduced compound. The valency states are CH_4, -4; CO_2, $+4$; NH_3, -3; NO_3^-, $+5$; SH_2, -2; and SO_4^{2-}, $+6$. There is, however, a very significant difference in the free energies obtained from the oxidation of the most reduced species of each as illustrated in Fig. 22. Each pair is separated by a line representing the $8\,e^-$ difference between them. The figure sums up many of the principles which were discussed in Chapter 1 with relation to chemotrophic bacteria

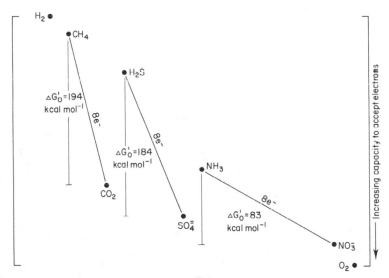

Fig. 22. A comparison between C, S and N oxidation/reductions. The most reduced and the most oxidized compounds of the C, S and N cycles are arranged in pairs, separated by a distance which represents an $8\ e^-$ difference between the extremes. Given vertically are the G'_o for the oxidation, by O_2, of the reduced form. There is a decreasing energy yield through the series C, S to N which is represented by the vertical distance between the oxidized and the reduced forms. The location of the lines relative to each other is only approximately correct and is designed to illustrate the decrease in reducing potential through the series H_2, CH_4, H_2S to NH_3 and the increase in oxidizing potential through the series CO_2, SO_4^{2-}, NO_3^- to O_2.

and their capacity to utilize for cell synthesis the energy from oxidation/ reduction reactions.

All three element cycles involve transitions from oxidized to reduced states and vice versa, but the carbon cycle differs from the other two cycles in being less involved with the completely reduced compound; methane is relatively unimportant in the main carbon flow as illustrated in Fig. 23. This figure shows the main pathways of carbon flow, particularly involving bacteria. The boxes represent the main carbon species plus hydrogen, they do not represent quantities. The width of the arrows represents an estimate of the relative importance of the pathway. The small flow into and out of methane may be compared with the large flux through CH_2O-carbon at the oxidation level of cellular material.

4.1. DISSIMILATIVE REACTIONS

Quantitatively, bacteria are much more important in dissimilative carbon metabolism than in carbon dioxide assimilation. The main input in the latter is from plant products. This, and the primary degradation of polymeric carbon compounds was discussed in Chapter 3. The rate of carbon cycling

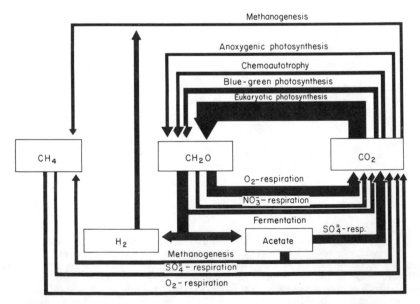

Fig. 23. The microbial carbon cycle. The role of sulfate in the oxidation of methane is largely hypothetical.

is primarily determined by the susceptibility of detritus to hydrolytic attack. We now look at the fate of the small molecular weight carbon compounds which are released as a result of that attack, considering in turn the individual pathways in Fig. 23.

4.1.1. Oxygen Respiration

The microorganisms which oxidize low molecular weight carbon compounds are principally those which are responsible for the initial hydrolytic attack. Their diversity lies in the range of degradative enzymes rather than in the mechanisms which they employ to process the small molecules for oxidative phosphorylation. The latter process involves the generation, through the Krebs (TCA) cycle, of reduced pyridine nucleotides, at the expense of acetate oxidation to carbon dioxide. The electrons from the NADH pass through a flavin and cytochrome chain, producing ATP. The final products of oxygen respiration are carbon dioxide plus cells, which are, in turn, partially broken down and oxidized when they die. We would like to know to what extent oxygen is involved in carbon oxidation in different ecosystems, but this is not always easy to determine.

Soil is predominantly an oxygen-rich environment and carbon oxidation is dominated by oxygen-respiring fungi and bacteria. The hydrolytic micro-organisms· to a large extent, utilize the hydrolytic products, but not exclusively. It is very commonly found that cellulose degradation proceeds more rapidly when non-cellulolytic bacteria are also present (Enebo, 1951). The exact mechanism of this mutualism is not known, but it is possible that efficient scavenging of hydrolytic end-products allows the cellulases to operate more efficiently and induces or derepresses the synthesis of more enzyme, thus causing a more rapid rate of cellulose hydrolysis.

The rates of decomposition and oxidation of various plant residues were seen in Fig. 18 and the factors influencing decomposition were discussed. Presumably, small anaerobic microniches may exist within an aerobic soil, as methane may be detected within soils (Bollag and Czlonkowski, 1973), and obligate anaerobes (*Clostridium* spp.) can easily be isolated from soils.

The extent to which aerobic decomposition in aqueous systems occurs, depends on the quantitative input of organic material, its particle size and rate of sedimentation, the depth of water through which it descends, the temperature of the system, and the degree of stratification and oxygenation of the water body. Deep oligotrophic lakes and oceans have a very low sedimentation rate since most of the carbon in detritus is oxidized as it sinks. Bacteria colonize sedimenting detritus (Wojtalik, 1970; Paerl, 1973), and up to 98 % algal cells can be decomposed in the epilimnium (Ohle, 1962).

Significant mineralization may, however, occur in marine sediments. Approximately 30% of the total pelagic nitrate is contributed by mineralization processes in the sediment underlying 4000 m Eastern Equatorial Atlantic waters (Bender *et al.*, 1977). This would indicate that an equivalent portion of carbon mineralization might be expected at the same sediment site. Many inshore sediments receive considerable quantities of carbon-rich detritus, and some larger bodies of water such as the Black Sea, the Cariaco Trench, and many fjords may have an oxygen-depleted bottom layer due to stratification and oxygen utilization by bacteria degrading organic matter. This also occurs in water, to which is added large quantities of untreated sewage or industrial waste, rich in carbon compounds. The rate of oxygen removal by microorganisms exceeds the rate of re-solution and diffusion to the deficient zone. This has undesirable effects on other forms of life in the water, and sewage treatment is usually necessary to eliminate the problem. In general, the treatment process consists of an aerobic flocculation/oxidation step, often followed by anaerobic digestion. The combined steps are analogous to natural sedimentation and sediment decomposition, but the organic content is much higher, the temperatures are higher, and the rates of decomposition are greatly accelerated. The kinetics of sewage digestion have been more thoroughly studied than those of sediment, and much of what is assumed to happen in sediments is deduced from a knowledge of sewage sludge and ruminant fermentations.

The upper strata of most sediments are oxic and decomposition of detritus by aerobic microorganisms continues there. This normally causes a depletion in oxygen, which can be replaced by diffusion and by turbulence, generated by local fauna or by waves. An oxygen gradient is generated with a reduced zone lying beneath it. There is a sharp change in redox potential at this point, which may be detected using a platinum electrode. In marine sediment systems this redox-discontinuity may be predicted from the rate of organic input and the hydrodynamic forces operating on the sediment (Fenchel and Riedl, 1970). The microbial activity in the oxic part of sediments may be studied by techniques which have proved useful in soil studies. The decomposition of ^{14}C-labeled glucose (Sorokin and Kadota, 1972), of ^{14}C-labeled barley straw (Fenchel and Harrison, 1976), of cellulose in litter bags (Hofsten and Edberg, 1972) provides useful information on microbial activity towards specific substrates while dehydrogenase activity (Lenhard, 1968) and ATP concentration (Holm-Hansen and Booth, 1966) yield more general information on bacterial activity and biomass. Perhaps the most useful method for estimating the total activity of sediments is to determine the rate of oxygen uptake (Pamatmat, 1971; Edberg and Hofsten, 1973; Jørgensen, 1977a). Oxygen consumption may be partly chemical and will also include the biological oxidation of some of the products of anaerobic processes from the underlying sediment: short chain fatty acids, hydrogen, hydrogen sulfide, ammonia, and methane.

The extent to which these compounds may enter the oxic zone will be quite different and will depend on the composition of the anoxic sediment; for example, a high sulfate content will have a considerable influence. The concentration of amino acids and sugars (Degens, 1970; Weiler, 1973; Degens and Mopper, 1975) is very much higher in marine sediments than in the overlying water, with steep gradients in the sediment. It seems likely that some of these hydrolytic products may diffuse along these gradients to an oxic zone. It is not easy to evaluate what proportion of the oxygen utilization is due to bacteria and what is due to eukaryotic forms either in soils or sediment. Bacteria are thought to be responsible for 80% of the oxygen demand in detritus in Lake Harwell (Abernathy, quoted by Kerr et al., 1972).

The oxidation of methane deserves special mention in relation to a reduced product of anaerobic metabolism reaching an oxic zone. This is because the relative insolubility of methane in water means that its arrival at an oxic site is not limited by diffusion. It often bubbles up through sediments to be oxidized in the upper sediments or the overlying water or to escape to the atmosphere. Methane-oxidizing bacteria have been isolated from a wide variety of locations (Whittenbury et al., 1975; Rudd and Hamilton, 1975). Methane-oxidizing bacteria are often located in a narrow zone within the thermocline where the oxygen concentration is low. This was thought to be a reflection of a microaerophillic metabolism, and this is true but in a rather specialized way. Rudd et al. (1976) showed that the methane oxidizers were nitrogen-depleted and relied on nitrogen-fixation to meet their growth demands. Nitrogen-fixation was only possible at low oxygen concentrations. During the winter months, when the lake turned over, nitrogen was no longer limiting, and methane oxidation was thought to be responsible for the major portion of oxygen depletion under frozen lakes.

4.1.2. Nitrate Respiration

It is generally thought that oxygen, nitrate, sulfate, and carbon dioxide are used in this order of preference as electron acceptors by mixed populations of bacteria (Mechalas, 1974). This concept is illustrated in Fig. 24, where it is seen that this order of utilization follows depth and successively decreasing availability of the preceding acceptor. It follows that after oxygen, nitrate is the acceptor of choice, this being a natural consequence of the decreased energy available from the oxidation of carbohydrate or hydrogen, when the different acceptors are utilized.

It is very likely that nitrate respiration (denitrification) can occur in an environment in which oxygen can be detected. Oxygen respiration and nitrate respiration are thus not completely exclusive. At dissolved oxygen concentrations of between 4 and 10 per cent saturation, ammonia in animal wastes was extensively nitrified and denitrified, when incubated in well-mixed chemostats (Smith et al., 1976). There was no accumulation of nitrate, and it must be presumed that nitrate competed with oxygen possibly in anoxic floccular microniches. This concept of anaerobic niches, located in an oxic environment is discussed further in Section 4.1.4.

Denitrification is extensive in soils which undergo flooding and thus become anaerobic. Presumably a similar situation exists in marine salt marshes and mud banks which are alternately flooded and exposed by tidal flow. Ammonia which is generated in the anoxic sediment would be nitrified when the sediment is exposed, and when covered again the nitrate may be denitrified at the expense of reduced carbon compounds, when the oxygen is depleted. A similar process might occur on a smaller scale in shallow non-tidal marine sediments where bacterial photosynthetic species are capable of rapidly oxidizing sulfide and thus allowing oxygen into the top of the sediment, when illuminated (Blackburn et al., 1975; Hansen et al., 1977). Sulfide would diffuse back into this layer at night and might itself as act a reductant for any nitrate that had accumulated. Such a process is discussed in relation to Thiobacillus denitrificans (Section 6.3).

The extent to which carbon is oxidized by nitrate in normal soils, marine and lake sediments is not very well known. There are considerable technical difficulties in measuring dinitrogen production in systems which already contain considerable quantities of this gas. This problem can partly be overcome in soils, by stripping the soil of its nitrogen and measuring the production from denitrification. This is not possible in a stratified aqueous sediment. The alternative is to use ^{15}N-labeled nitrate and to measure $^{15}N_2$ production, procedure used by some investigators (Goering and Dugdale, 1966a, b; Chen et al., 1972; Koike et al., 1972). Unfortunately, most experiments were performed using unnaturally high nitrate concentrations (Section 5.3.2).

As the rates of denitrification are usually not precisely known, it follows that the quantities of carbon oxidized by nitrate reduction are also unknown. An added uncertainty is introduced into any calculations by a lack of knowledge regarding the products of denitrification and the extent to which nitrite is denitrified. Depending on these factors, 2, 3, 4 or 5 carbon atoms can be oxidized for every 4 nitrogen atoms reduced. It is, therefore, not easy to use denitrification as an index of carbon oxidation nor to calculate denitrification from carbon dioxide production. It is, however, of interest to determine to what extent carbon oxidation might proceed through nitrate reduction. It is necessary to know the carbon flow through other oxidants and for this reason it is useful to consider an example of a well analyzed marine sediment where these rates have been determined. Some data are summarized in

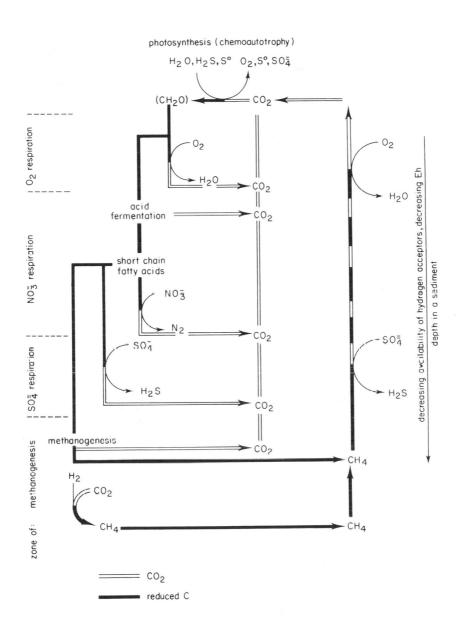

Fig. 24. The successive involvement of oxygen, nitrate, sulfate and carbon dioxide as electron acceptors with decreasing Eh.

Table IX, where all processes involving the aerobic and anaerobic respiration and assimilation of carbon are given. The rates of nitrate reduction (Section 5.3.2) are at best approximations but it is unlikely that they could exceed the upper value, which would make a very small contribution (2%) to the total carbon respiration. For comparison, the range of denitrification rates quoted (Section 5.3.2) of 0.2–72 mmol N m^{-2} day^{-1} would result in carbon oxidation rates of 0.25–90 mmol C m^{-2} day^{-1} and carbon assimilation rates of 0.17–60 mmol C m^{-2} day^{-1}.

A very different situation can be found in marine environments, deficient in oxygen but rich in nitrate. In Darwin Bay, Galapagos Archipelago, rates of nitrate reduction to nitrite of 0.16μmol l^{-1} day^{-1} for water below 40 m and of 3.56 mmol m^{-2} day^{-1} for sediments were found (Richards and Broenkow, 1971). These rates of nitrate reduction could account for an oxidation of 50% of the organic carbon produced in the water column. This value for nitrate reduction in mmol m^{-2} day^{-1}, of 3.65 is more similar to the range of 0.26–5.19 (calculated in Section 5.3.2) for the Ostend Sluice Dock (water nitrate = 100μM) than for the Limfjord sediments (water nitrate = 0–15μM). Similarly, higher rates of carbon oxidation would be anticipated for the Ostend sediments. The factors which regulate denitrification and coupled oxidation in soils are so complex that no attempt will be made to quantify carbon flow through this pathway, but it is likely to be low except in unusual circumstances.

4.1.3. Fermentation

The fermentative bacterial processes were discussed in some detail in Section 1.1, and in the absence of further information relating to the types of bacteria and even the final products of fermentation in lake or marine sediments, it may be assumed that the processes are similar to, but slower than, ruminant and sludge fermentation (Section 2.1.1). The final products, as a result of coupled fermentations and hydrogen utilization, are the short chain fatty acids, principally acetate and some hydrogenated compound (methane or hydrogen sulfide). The intermediate and final products of fermentation depend very considerably on the material being fermented and the microbial population present. This has been demonstrated for rumen fermentation where the ratio of acetic/propionic acid is higher from alpha-cellulose and hemicellulose when fermented by an inoculum from a hay-fed animal than by an inoculum from a hay–grass-fed animal and that the two substrates themselves yield different ratios (Satter et al., 1964). Hungate (1966) discusses the stoichiometry of ruminant fermentations with relation to substrate composition (see Section 8.1). The residence time is relatively short in ruminants, and this has a marked effect on the fermentation products. This

TABLE IX. Respiration and Assimilation of Carbon in Limfjord Sediment.

Reductant	Oxidant	Oxidant consumed (mmol m^{-2} day^{-1})	Carbon oxidized (mmol m^{-2} day^{-1})	Ratio C_{inc}/C_{ox}	Carbon incorporated (mmol m^{-2} day^{-1})
CH_2O	O_2	17[a]	17 (46%)	1·00	17 (75%)
1·5 CH_2O	NO_3^-	0·01–0·8	0·01–0·60 (2%)[b]	0·67	0·01–0·40 (2%)
2 CH_2O	SO_4^{2-}	9·5	19 (52%)	0·28[c]	5·23 (23%)
Total			36·6		22·6

[a] Data from Jørgensen (1977a).
[b] The carbon oxidized by nitrate is calculated on the basis of complete reduction of nitrate to dinitrogen. Other rates and calculations are discussed in the text.
[c] Based on 69 lactate-C oxidized for every 19 C incorporated (see Table XII).

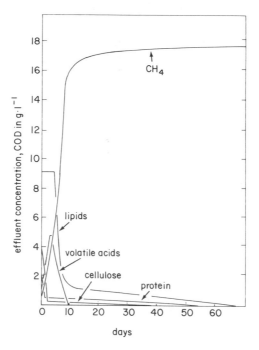

Fig. 25. The effect of retention time on the relative breakdown at 35°C of protein, cellulose, lipids and their fermentation products measured as oxidation equivalents and which are mainly the volatile short chain fatty acids, acetate, propionate and butyrate. Methane increases with retention time at the expense of the volatile acids (redrawn from Lawrence, 1971).

point is illustrated in Fig. 25. No distinction can be drawn between carbo-hydrates and protein digestion from this data, but the conversion of the volatile acid fermentation products acetate, propionate and butyrate, to methane, as a function of residence time is clearly demonstrated.

The carboxylic acids, lactic and succinic acids, do not accumulate in these rumen or sludge fermentations, even though they are intermediates. Lactic acid can be a significant fermentation intermediate in the rumen of starch-fed animals, and is thought to be further fermented to propionate (Walker, 1968). Succinic acid which occurs in very low concentration in the rumen, may be the major precursor of propionate as it has such a rapid turnover (Blackburn and Hungate, 1963). This illustrates a general principle that the quantitative importance of an intermediate is very often not correlated with its concentra-tion in a system; the more important it is, the faster is its turnover and the lower is its concentration.

The concentration and turnover of intermediate fermentation products has not been extensively investigated in sediment systems. Cappenberg and

Prins (1974) found turnover rates of L-lactate and acetate to be 28·9 and 2·4 μg/g of mud per hour, respectively. The relatively very high turnover of lactate in this system (Lake Vechten sediment) is unusual, and is probably not representative of other ecosystems. The normal product of further lactate fermentation would be propionate, which itself would be fermented to acetate, as would butyrate (Pine, 1971; Bryant, 1976).

The involvement of sulfate reduction and methane production with the fermentation process and the oxidation of the end-products of fermentation will be the topic of the next two sections.

4.1.4. Sulfate Respiration in Sediments

Sulfate plays a role analogous to nitrate, but since the energy yield with sulfate as an electron acceptor is less than with nitrate, the latter is used preferentially. Systems containing aggregated carbon might be expected to show both nitrate and sulfate respirations. Oxygen respiration and sulfate respiration can occur in the same sediment stratum; Jørgensen (1977c) has clearly shown sulfate reduction in anoxic microniches in oxic marine sediments. Jørgensen has calculated that in oxygen-saturated seawater, the particle size would have to be greater than 2 mm to be anaerobic at the center or 200 μm at 1% oxygen saturation. The contribution of these microniches to organic decomposition may be considerable, as can be judged from the high rate of sulfate reduction/carbon oxidation in the upper, oxic layers. The oxidizing capacity of sulfate in seawater (20–30 mM) is 200 times that of oxygen-saturated seawater (0·2–0·3mM), since one sulfate is equivalent to two oxygen molecules in oxidizing power. Sulfate is the predominant electron acceptor in upper, anoxic, marine sediments. This has been quantitatively demonstrated by Jørgensen (1977a) who has shown sulfate reduction rates of 25–200 nmol S reduced cm^{-3} day^{-1}, comparable to the 30–1200 nmol S reduced cm^{-3} day^{-1} quoted by Goldhaber and Kaplan (1974). It corresponds to an average rate of 9·5 mmol S m^{-2} day^{-1}, this being a total of the sulfate reduction in the underlying layers. This sulfate reduction is equivalent to an oxidation of 19 mmol C m^{-2} day^{-1}. A separate measurement of oxygen uptake by the sediment surface indicated 34 mmol O$_2$ m^{-2} day^{-1} consumed. At least 50% of this oxygen consumption was in oxidizing the sulfide which reached the surface, thus only 17 mmol C m^{-2} day^{-1} could have been oxidized by oxygen, compared to 19 mmol C m^{-2} day^{-1} oxidized by sulfate.

These results for sulfate and oxygen respiration are summarized in Table IX, for comparison with nitrate respiration, which is seen to be a minor contribution to carbon oxidation in this system. The greater efficiency of oxygen-respiring bacteria is reflected in their greater contribution (75%) to the cells synthesized by respiratory processes. Sulfate reduction alone (the efficiency of 0·31 included fermentation) is relatively inefficient, accounting for only 23% of cell synthesis, nitrate-respiring bacteria contributing the balance of 2%.

The bacteria capable of oxidizing reduced compounds by sulfate in energy-yielding reactions are discussed in Section 1.1.1. Lactate, acetate, hydrogen, and methane are oxidized, yielding hydrogen sulfide, carbon dioxide, and water as products. The relative proportions of the compounds which are oxidized is not known, but all the reactions are biologically significant. The quantity of lactate turned over by this process is perhaps the least significant, although until recently lactate was the only carbon compound of note, known to be oxidized by sulfate. The oxidation of hydrogen by sulfate has long been known, but the significance in promoting hydrogen-producing fermentations is only now being appreciated. This aspect of fermentation by species combining together is discussed in Section 1.1.1, where it was emphasized that unless molecular hydrogen was removed by oxidation with sulfate or carbon dioxide, fermentation was inhibited. Acetate is an end product of many fermentations, and the capacity of bacteria to oxidize it with sulfate is most important. It is possible that a major portion of the carbon flow is through this process.

Methane is not found in sulfate-rich sediments (Fig. 26), and it has been thought that sulfate may inhibit methane production, and methane may re-

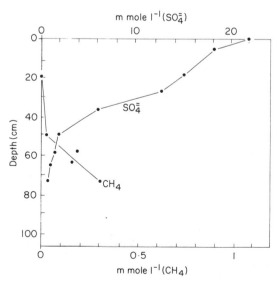

Fig. 26. Sulfate and methane profiles in the interstitial water of Long Island Sound sediments (redrawn from Martens and Berner, 1974).

duce sulfate (Martens and Berner, 1974). The latter suggestion is probably true but there is evidence for methane production and consumption in the same environment (Barnes and Goldberg, 1976). This will be discussed further in Section 4.1.5; the importance of the methane/sulfate oxidation is that little methane escapes through the sulfate-rich upper sediments, except by bubble transport (Martens, 1976). Presumably methane oxidation must account for a large portion of sulfate reduction in the lower sediments. Some aspects of this connection between the sulfur and carbon cycles are illustrated in Fig. 27. One of the main results of coupled hydrolysis, fermentation, and sulfate reduction is the maintenance of a neutral pH due to the liberation of ammonia, which is not oxidized in anoxic environments.

The effect is to convert bicarbonate to carbonate, which precipitates as calcium carbonate. An equally significant consequence of sulfate reduction is the transfer of reducing power from detritus to soluble ammonia and hydrogen sulfide, which can then diffuse to the upper sediment and act as energy sources for chemolithotrophic bacteria.

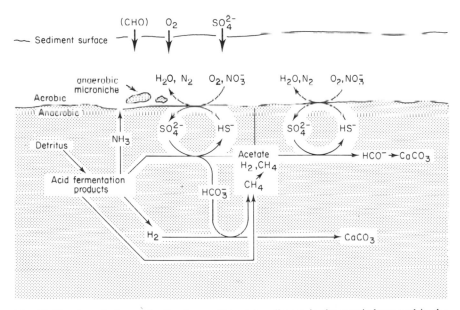

Fig. 27. The degradation of detritus in a sulfate-rich sediment, in the anoxic layer and in the anoxic microniches. Detritus is hydrolyzed and fermented, liberating ammonia. The oxidation of fermentation products (e.g., hydrogen and lactate) are coupled to sulfate reduction by *Desulfovibrio* producing acetate, or in the case of hydrogen, to methane production by methanogenic bacteria. Acetate, hydrogen and methane can be further oxidized by *Desulfovibrio* or *Desulfotomaculum* species, producing carbon dioxide and hydrogen sulfide. A combination of these processes leads to carbonate precipitation. The hydrogen sulfide can diffuse to an oxic zone and be oxidized at the expense of oxygen or nitrate.

The capability of the fermenting bacteria, coupled with the sulfate reducers, to mineralize dead bacterial cells and at the same time maintain a pH suitable for growth, is important in creating conditions, where further growth and mineralization can proceed in partially closed systems, without the input of further nutrients.

4.1.5. Carbon Dioxide Respiration, Methanogenesis

The major precursors for methane production are acetate and hydrogen/carbon dioxide (Fig. 27). Acetate is the precursor for 70% of methane produced in sewage sludge (Smith and Mah, 1966), for 60% of the methane from rice paddy soils (Koyama, 1963), and for 75% of the methane in Lake Vechten sediments (Cappenberg, 1974; Cappenberg and Prins, 1974). Acetate was shown to be a major intermediate in benzoate degradation to methane by a consortium of bacteria (Ferry and Wolfe, 1976), yet isolates of acetate-utilizing methanogens could not be made from this consortium nor from many other enrichments by other investigators. Cappenberg (1974), however, isolated a *Methanobacterium* sp. from lake sediment, which used acetate. *Methanosarcina bakeri* is one of the best documented species, which has been isolated and shown to produce methane from acetate at a good rate (Mah *et al.*, 1976). Increased rates of methane production were obtained when the medium was supplemented with trypticase and yeast extract, or when a number of other bacteria were present, presumably supplying growth factors and nutrients, or removing inhibitors. Zeikus *et al.* (1975) were unable to obtain methane from cultures of *M. bakeri* or *Methanobacterium thermoautotrophicum* which were supplied with acetate, unless also supplied with hydrogen. They speculate that the $\Delta G'_o = -6 \cdot 71$ kcal/mol is insufficient to allow for ATP generation. In general it is the methyl group of acetate which is converted to methane, but these workers found that both carboxyl and methyl carbons went to methane as might be expected from:

$$CH_3COO^- + 4H_2 \rightarrow 2CH_4 + 2H_2O, \qquad \Delta G'_o = -39 \text{ kcal.}$$

Cappenberg and Prins (1974) found that only 13% of the carboxyl carbon was converted to methane, and concluded that carbon dioxide was an intermediate. It is, therefore, not entirely clear what the role of acetate is in natural systems, but it would be safe to say that it plays a major role in methanogenesis, and it has been suggested (Mah *et al.*, 1976) that clostridia may utilize carbon dioxide plus hydrogen to first form acetate, in preference to its use by methanogenic bacteria to form methane directly.

The importance of hydrogen utilization by methanogenic bacteria has been stressed in previous sections, as it allows many hydrogen-producing fermentations to proceed, which would otherwise be inhibited by molecular hydrogen (Bryant et al., 1967; Scheifinger et al., 1975; Wolin, 1974).

The utilization of methane and acetate by sulfate-reducing bacteria and the thermodynamic advantage of utilizing sulfate rather than carbon dioxide as an electron acceptor have already been discussed in this chapter. It is thus not surprising that little methane is found in sulfate-rich sediments or seawaters. There is possibly some methane production in sulfate systems (Barnes and Goldberg, 1976) just as sulfate and nitrate reduction can occur in oxygen environments. This may again be due to microniches, more reduced than their surroundings. Oremland (1975) demonstrated methane production in rich organic, shallow-water, tropical marine sediments and postulated a regulation by photosynthetic oxygen production. It would seem likely that methane will only accumulate in shallow sediments which are rich in organic material.

A pure culture of a *Methanobacterium* sp. was completely inhibited by hydrogen sulfide 0·1 mM but was unaffected by 0·001 mM (Cappenberg, 1975). A pore water concentration of sulfide up to 0·28 mM in Lake Mendota sediment caused no inhibition, however, and even at 25 mM sulfide, methanogenesis was not completely inhibited (Winfrey and Zeikus, 1977). It is not known what role the high sulfide content of marine sediments may play in the regulation of methane production in these systems.

Methane production in salt marsh sediments and in whole-cell suspensions of *Methanobacterium thermoautotrophicum* and *Methanobacterium formicicum* was inhibited by nitrate, nitrite, nitric oxide and nitrous oxide (Balderston and Payne, 1974). These nitrogen oxides also inhibit methanogenesis in soils (Bollag and Czlonkowski, 1973). The nitrogen oxides, in order of effectiveness in inhibiting methanogenesis in salt marsh sediments, were nitric oxide, nitrous oxide, nitrite and nitrate. This order, with nitrate being the least effective, together with data on oxygen sensitivity and substrate excess, led Balderston and Payne (1976) to conclude that the inhibition was not due to either a change towards a more oxidized environment nor to a competition between nitrate- and carbon dioxide-using processes. The mechanism of inhibition is not understood.

Methane production can proceed rapidly at 60°C (Bryant et al., 1976); sewage sludge digestors operate at 35–40°C and methane production was observed in lake sediments between 4 and 40°C (Zeikus and Winfrey, 1976). A seasonal drop in temperature of 12°C from a maximum of 23°C resulted in a 100- to 400-fold decrease in the rate of methane production. It would seem that low temperatures do not select for methanogenic bacteria with low temperature optima.

Methane production in sewage sludge digestion is sensitive to excess acid production which causes a fall in pH (Kugelman and Chin, 1971), methanogenic bacteria can only grow within the narrow range of pH 6·4–7·5. They are able to maintain this pH by converting acids to methane but if the rate of acid production exceeds the rate of removal, resulting in a pH fall, the system becomes "stuck" or "sour" and cannot recover without external help. Possibly this may happen in natural environments also; some bogs and mires may be examples of such "sour" fermentations (Section 3.4).

The formation of methane represents a mechanism which enables mineralization of sedimented carbon to proceed in sediments in the absence of oxygen, nitrate or sulfate. In sulfate-deficient lakes it probably is the most important pathway for carbon mineralization. In Lake Vechten a large proportion of mineralized carbon flows through the methane pool (Cappenberg, personal communication). This is also true of marine deposits under 2 m deep, where sulfate often becomes limiting (Reeburgh and Heggie, 1974). Claypool and Kapan (1974) calculate that a minimum of 20 mmol $CH_4 l^{-1}$ interstitial water is formed in deep sea sediments which receive a rapid input of detritus. This quantity of methane is close to the amount necessary for gas hydrate formation, if there is sufficient (500 m) hydrostatic pressure and a temperature of 5°C. The location and formation of gas hydrates is extensively discussed by a number of authors (Kaplan, 1974).

When methane bubble formation does occur, there is a preferential stripping of nitrogen from sediments compared to argon, resulting in a lowered nitrogen to argon ratio (Reeburgh and Heggie, 1974).

4.2.   AUTOTROPHIC REDUCTION OF CARBON DIOXIDE

The contribution of prokaryotic carbon dioxide reduction to primary productivity is small compared to plants, and is principally due to blue-green bacteria. It is doubtful if carbon dioxide fixation by anaerobic photosynthesis or the chemolithotrophic autotrophs can be described as "primary productivity" as both processes require the oxidation of reduced inorganic compounds, such as hydrogen sulfide, which probably owes its reduced state to the reduction of sulfate by a plant tissue. It is possible, however, to run a sealed anoxic microcosm, driven by light, in which only anaerobic photosynthesis operates, supplying cell products for the reduction of sulfate. In real situations these definitions are irrelevant, the important fact is that carbon dioxide fixation can occur through a bacterial system, utilizing products whose potential energy would otherwise be lost to the biosphere. The resulting cellular carbon can enter food chains for the support of higher forms of life.

4.2.1.   Blue-Green Bacteria

The blue-green bacteria have a widespread distribution (Carr and Whitton, 1973; Whitton and Sinclair, 1975), but on a global basis their contribution to total primary productivity cannot be large. They are, however, very important in specific habitats. The factors which make the blue-greens particularly successful in these habitats will be briefly discussed.

Carbon dioxide concentration. It is a general characteristic of blue-green bacteria that they are more efficient than green algae in utilizing low concentrations of dissolved carbon dioxide (Shapiro, 1973). Associated with this property is their competitive advantage over green algae at high pH values, where free carbon dioxide is at a lower concentration. Perhaps the blue-greens have sacrificed some versatility in achieving this efficiency due to acid-sensitive chlorophyll (Brock, 1973a). *Nostoc commune* completely dominates limestone pools (Whitton, 1971). Possibly this capacity to grow at low carbon dioxide tensions is associated with the ability of blue-greens to multiply in surface flocs, where little carbon dioxide can be available.

Oxygen concentration. Blue-greens may often be inhibited by high oxygen concentrations, particularly when illuminated (Lex *et al.*, 1972). They appear to be particularly sensitive to photooxidation when deprived of carbon dioxide (Abeliovich and Shilo, 1972). The ability to grow at low oxygen tensions has a marked effect on the distribution of the blue-greens. They are often located in near-anaerobic paddy fields, estuarine muds and thermal springs and may even utilize hydrogen sulfide as an electron donor (Cohen *et al.*, 1975). They are able to survive in dense flocs where the oxygen tension is low.

Temperature tolerance. Blue-green bacteria are capable of growth up to 75°C, whereas green algae cannot grow above 56°C and are not competitive with the blue-greens in alkaline waters above 40°C (Brock, 1973b).

Gas vacuoles. Many blue-green bacteria possess gas vacuoles which allow them to rise through the water, sometimes to the surface, to a point where they can obtain light for photosynthesis. The gas vacuoles probably collapse, due to photosynthetic activity, allowing the bacteria to sink again. This passage through the water column enables them to obtain nutrients from a considerable volume of water. Flocs are formed when large numbers of floating cells are collected by surface currents. Blue-greens are more commonly observed in eutrophic than in oligotrophic waters (Whitton, 1973) but they may be present unobserved in the depths of oligotrophic lakes. The contribution of blue-greens to productivity in the oceans cannot be assessed.

Nitrogen fixation. The capacity of some blue-greens to reduce dinitrogen is important in determining their distribution in nitrogen-deficient environments.

4.2.2. Anaerobic Photoautotrophs

The range of diversity in this group of bacteria is discussed in Section 1.2. Their distribution is limited by their requirement for both light and reduced inorganic compounds, which are only produced in anoxic conditions. These bacteria are thus limited to growth on anoxic sediment surfaces covered by shallow water or in stratified lakes and fjords. They are not of major interest in the context of carbon cycling. Some examples of carbon dioxide fixation, associated with sulfur oxidations, are discussed in Section 6.4.2.

4.2.3. Chemoautotrophs

These bacteria are dependent on reduced inorganic substrates for energy and with the exception of ammonia these substrates are exclusively the products of anaerobic metabolism. The bacteria are thus generally located at anaerobic/aerobic interfaces. Unlike the sulfide-phototrophs they can be located below the sediment surface, thus extending their range of habitats.

The oxidation and incorporation of methane while not being a strictly autotrophic process, has autotrophic features and in a practical sense prevents the loss of a considerable quantity of carbon from the biosphere. These bacteria can incorporate into cells one-third of the methane taken up, two-thirds being converted to carbon dioxide (Rudd *et al.*, 1974). In sulfate-deficient waters and sediments, the considerable mineralization of organic carbon to methane would indicate that a good yield of bacterial cells could be produced from this methane. Methane can be oxidized at the rate of 240 μmol m^{-3} day^{-1} at 5°C (Rudd *et al.*, 1974) which at one third incorporation could result in the cellular incorporation of 0.08 mmol C m^{-2} day^{-1}, if the methane oxidizers were stratified over one vertical meter.

The oxidation of ammonia to nitrate yields a small quantity of cells (Table X), the overall maximum incorporation is 1 mol bicarbonate/5 mol ammonia oxidized (Gundersen and Mountain, 1973). Using this or more complex formula (Knowles *et al.*, 1965) the cell yield can be calculated from the rate of ammonia oxidation. The possible contribution of nitrification to carbon dioxide reduction in Limfjord sediments was calculated to be only 0.02–0.04 mmol C m^{-2} day^{-1} (Table X).

The corresponding carbon incorporation for sulfide oxidation is much higher, 5.13 mmol C m^{-2} day^{-1}. The assumption is made that *Thiobacilli*-like bacteria may be involved in the process, but in reality *Beggiatoa* species are the predominant microorganisms (Jørgensen, 1977c). As there is considerable doubt concerning the ability of these bacteria to utilize the energy gained from the oxidation of hydrogen sulfide or to reduce carbon dioxide, the calculation in Table X is a little doubtful. There is no doubt, however,

TABLE X. Contribution of Lithotrophic Bacteria to CO_2 Incorporation in Limfjord Sediment.

Reductant	Oxidant	Reductant oxidized (mmol m^{-2} day^{-1})	Molar Ratio[b]	C incorporated (mmol m^{-2} day^{-1})
H_2S	$2O_2$	19[a]	0.27[c]	5.13
NH_4^+	$2O_2$	0.1–0.2	0.20[d]	0.02–0.04
Total				5.16

[a] Data from Jørgensen (1977a).
[b] Carbon incorporated/reductant oxidized.
[c] The assumption is that thiobacilli or bacteria with a similar efficiency (energy used in CO_2 reduction/total free energy available) is 20% (Larsen, 1960). The energy for CO_2 reduction is 118 kcal, the energy from the oxidation of H_2S to SO_4^{2-} is 160 kcal (Sokolova and Karavaiko, 1964).
[d] Based on Gundersen and Mountain (1973).

that *Beggiatoa* species constitute 18–71 mmol C m^{-2}, assuming a 10:1 biomass to dry weight ratio and 50% carbon content (Jørgensen, 1977c). It also seems possible that they have used the plentiful H_2S energy and CO_2–C in generating this biomass, as other energy and carbon sources are limited.

4.2.4. Carbon Budget in a Sediment

The phytoplankton productivity in the Limfjord waters is 100–200 g C m^{-2} year^{-1} or 20–40 mmol C m^{-2} day^{-1}. The total oxidation of carbon (Table IX) is 36 mmol C m^{-2} day^{-1}, thus approaching the maximum phytoplankton input. Associated with this carbon oxidation is a calculated heterotrophic assimilation of organic carbon into cells of 22.6 mmol C m^{-2} day^{-1} and of lithotrophic autotrophic cells of 5.2 mmol C m^{-2} day^{-1} (Table X). These cells themselves will die and be partially mineralized, some resistant cellular components will accumulate, slowly changing the sediment composition, the older lower portions becoming less degradable.

The carbon input from benthic photosynthesis and from eel grass is unknown but thought to be significant. The conclusion must be drawn that this type of sediment is very dynamic and that much of the primary productivity of the water must enter the sediments to maintain the observed high respiration rates.

The results from Jørgensen (1977a) indicate that the rate of carbon oxidation in the sediment would exhaust the carbon pool in 3–5 years. He draws the reasonable conclusion that fresh supplies of carbon must enter the lower layers of the sediment and that vertical mixing goes down to 30–40 cm.

4.3. THE PRECIPITATION AND DISSOLUTION
OF CARBONATES

One more aspect of the carbon cycle to be considered in this chapter is the state of the most oxidized form of carbon. In water it may occur as dissolved CO_2, as H_2CO_3, as HCO_3^- and as CO_3^{2-} which again coexists with precipitated carbonates. Essentially by taking up or releasing CO_2 or by influencing the hydrogen ion activity in other ways, living organisms influence the bicarbonate–carbonate system of natural aquatic environments.

The deposition of $CaCO_3$ (as calcite or aragonite) and $MgCO_3$ is often biologically mediated as is evident from the skeletons of corals, molluscs, foraminifera and other animals and algae. In photosynthesis, aquatic plants utilize CO_2, thus driving the reaction,

$$2HCO_3^- \rightarrow CO_2 + CO_3^{2-} + H_2O,$$

which again results in an increased value of pH. In waters saturated with $CaCO_3$, e.g., in some types of lakes and in tropical seas, this will lead to the precipitation of calcite. This process is also mediated by cyanobacteria which play a role in the $CaCO_3$ deposition in the form of stromatolithic structures in extant coral reefs (Sharp, 1969). The wide distribution of fossil stromatolites from the Precambrian shows that this earliest known biogenic $CaCO_3$ deposition mediated by blue-greens must have once played a very significant quantitative role (see Section 2.2).

Other bacterial processes may lead to the precipitation of $CaCO_3$. In evaporites, Ca^{2+} is precipitated as gypsum ($CaSO_4 \cdot 2H_2O$) and some other minerals containing SO_4^{2-} and Ca and Mg. When sulfate reducing bacteria are active in such environments they utilize the sulfate and the Ca^{2+} precipitates as carbonates. The process has been shown to be important in evaporites along the Red Sea. In "Solar Pond", a hyperhaline pond in partial connection with the Red Sea, Jørgensen and Cohen (1977) could show a deposition rate of $CaCO_3$ which quantitatively corresponded to the rate of sulfate reduction. The process of $CaSO_4$ dissolution and precipitation of $CaCO_3$ seems also to be of geochemical significance where oil deposits are in connection with evaporites (Ivanov, 1968). The problem of $CaCO_3$ deposition by sulfate reducers is further discussed by Friedman (1972) and by Deelman (1975).

Methane producers may contribute to the precipitation of carbonates, viz.,

$$Ca^{2+} + 2HCO_3^- + 2H_2 \rightarrow CaCO_3 + CH_4 + H_2O.$$

The formation of carbonate cements found in some sediment cores, taken in off-shore sediments, have been explained by this process (Allen et al., 1969; Hathaway and Degens, 1969).

Various heterotrophic bacteria are also known to catalyze the precipitation of carbonates in their environment or in bacterial slimes. The quantitative importance of this in, e.g., soils, is not known however. DiSalvo (1973), discussing the role of bacteria in forming limestone cements in sediments of coral reefs, emphasized ammonification, and thus an increase of pH as the important mechanism.

Microbial activity can also lead to the dissolution of carbonates. Acid-producing bacteria are probably important in the dissolution of carbonate and other (e.g., phosphate) minerals in soils. The oxidation of sulfides and sulfur by thiobacilli is a very acid-producing process (see Section 7.1.1). The sulfuric acid formed will attack carbonate rocks leading to the precipitation of sulfates. The importance of this process in connection with sulfur deposits is described in Ivanov (1968).

4.4. EQUILIBRIUM DISTRIBUTION AND DIAGENESIS

Organic matter is thermodynamically unstable and therefore decomposes into small molecules whose relative concentrations, at equilibrium, may be predicted (Thorstenson, 1970). Bacteria are largely responsible for the decomposition process but the equilibrium is independent of their activities and would be the same if the process were simply physicochemical. The equilibrium will depend on the composition of the organic material being decomposed and on the Eh and pH of the system. Diagrams may be prepared showing the predicted relative distributions of the predominant small molecules (CH_4, H_2CO_3, HCO_3^-, CO_3^{2-}, NH_4^+, NH_2OH, N_2, NO_3^-, H_2S, HS^- and SO_4^{2-}) in relation to pH and Eh (Thorstenson, 1970; Berner, 1971). These calculated distributions, at specified pH and Eh values are very similar to those observed in sediments.

Berner (1974) describes organic decomposition as being a first order process:

$$\frac{dG}{dt} = -kG,$$

where k = first order rate constant,
$\quad G$ = concentration of originally deposited decomposable organic material as mass per unit volume of pore water,
$\quad t$ = time.

An example of this type of decomposition as it occurs in upper marine sediments is quoted by Berner:

Proteins (A) $\longrightarrow$ Amino acids (A) $\longrightarrow$ NH$_3$

Proteins (B) $\longrightarrow$ Amino acids (B) $\xrightarrow{\text{SO}_4^{2-}}$ NH$_3$ + H$_2$S + HCO$_3^-$

Carbohydrate $\longrightarrow$ Sugars $\xrightarrow{\hspace{1.5cm}\text{SO}_4^{2-}\hspace{1.5cm}}$ H$_2$S + HCO$_3^-$

The G for ammonia production (G_N) would be the total decomposable protein ($A + B$), whereas the G for sulfate reduction (G_S), would be the decomposable protein (B) plus the decomposable carbohydrate. As an example of how the rate of ammonia production may be derived:

$$\frac{dC}{dt_{\text{biol}}} = -\alpha_N \frac{dG_N}{dt} = \alpha_N k_N G_N,$$

where α_N = mol ratio of nitrogen to carbon in the substrate,

$\quad\quad k_N$ = first order rate constant for ammonia production.

This may be substituted into some form of Berner's (1971, 1974) general diagenetic equations, which at its simplest has the following form:

$$\frac{\partial C}{\partial t} = D \frac{\partial^2 C}{\partial x^2} + \omega \frac{\partial C}{\partial x} - K \frac{dC}{dt} + \frac{dC}{dt_{\text{biol}}} - \frac{dC}{dt_{\text{min}}},$$

$$\underset{\text{Diffusion}}{} \quad \underset{\substack{\text{Sediment-}\\\text{ation}}}{} \quad \underset{\substack{\text{Ion exchange,}\\\text{adsorption}}}{} \quad \underset{\substack{\text{Biological pro-}\\\text{duction rate}}}{} \quad \underset{\substack{\text{Mineral pre-}\\\text{cipitation and}\\\text{solution}}}{}$$

where C = concentration of dissolved species, in mass per unit volume pore water,

$\quad x$ = depth measured positively downward from the sediment–water interface,

$\quad K$ = equilibrium constant times the ratio of the dissolved species undergoing ion exchange,

$\quad D$ = diffusion coefficient of the dissolved species in the sediment,

$\quad \omega$ = rate of sedimentation.

The reviews of Berner (1971, 1974) are recommended for examples of the application of the diagenetic equation to the computation of rates of hydrogen sulfide and ammonia production from their respective concentrations in marine sediments when values for the other factors in the equation are known.

CHAPTER 5

The Nitrogen Cycle

The main components of the nitrogen cycle are outlined in Fig. 28. In common with the sulfur cycle, it involves an eight electron shift between the most oxidized form (nitrate) and the most reduced form (ammonia), which like sulfide is the form in which it is incorporated into most biological materials. Unlike sulfate in the sulfur cycle, nitrate is not the predominant nitrogen compound in oxic environments, although thermodynamically it should be. Instead, gaseous dinitrogen is the largest nitrogen source, discounting the enormous quantities locked in sedimentary and igneous rocks as ammonia.

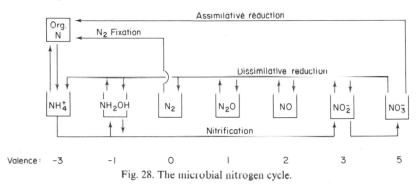

Fig. 28. The microbial nitrogen cycle.

The processes in the cycle will be discussed individually in detail so a review of the cycle will not be given at this stage. It is important, however, to emphasize that the major flow of nitrogen is from organic nitrogen to mineral products and back again, and small losses from nitrate to dinitrogen are matched by the process of dinitrogen reduction (nitrogen-fixation) to ammonia and subsequently to organic nitrogen.

5.1. AMMONIA LIBERATION AND REINCORPORATION

Nitrogen is present in cells principally in proteins and polynucleotides. When these compounds are hydrolyzed and catabolized, the nitrogen is

101

liberated in the form of ammonia. As ammonia is the end-product of animal excreta also, virtually all organic nitrogen breakdown flows through this compound. Quantitatively the yearly flow is enormous, being equivalent to almost the total productivity in oceans and on land. These systems, terrestrial and oceanic, are essentially closed in that most of the degradation products are reassimilated within each system and there is relatively little exchange between the two systems and relatively little exchange with the atmosphere.

The degradation of nitrogen in detritus is very similar to that of carbon; this has been discussed in Chapter 3. The conclusion that may be drawn from that chapter is that while pools of organic material exist in soil and in the oceans, these are turning over and there is almost no net accumulation. This is certainly true for soils but sedimentation in the oceans does account for a rate of 38 million tons N year^{-1}, a small amount compared to the total turnover and the enormous organic reservoir in soils of 3×10^5 million tons and of $5 \cdot 3 \times 10^5$ million tons of dissolved organics in the oceans (Söderlund and Svensson, 1976).

There are major differences between oceanic and terrestrial systems in the fate of ammonia which is liberated. As already stated, both systems are essentially closed and isolated from each other but some ammonia can enter the atmosphere from animal feces in grazed pastures (Denmead *et al.*, 1974) and half the ammonia from manure may escape to the atmosphere (Eriksson, 1959). A total of 20 to 50 million tons ammonia N year^{-1} may enter the atmosphere from animal and human wastes, a relatively small quantity compared to the 93 to 209 million tons ammonia N year^{-1} lost over the land due to a variety of largely unknown sources (Söderlund and Svensson, 1976). Much of this ammonia is reprecipitated on the land but a portion (19 to 50 million tons N year^{-1}) is deposited in the sea and a portion (3 to 8 million tons N year^{-1}) goes to oxides of nitrogen in the atmosphere.

Nitrogen mineralization (ammonia liberation) may occur in oxic or anoxic environments, the former represented by most soils and aerobic water bodies, the latter by anoxic oceanic and lake sediments. Some of the differences between the two types of environments are illustrated in Fig. 29. In both systems organic nitrogen is converted to ammonia which can exchange with cation exchange sites on soil particles, but then the differences become marked. The ammonia may become oxidized (nitrification) in the oxic soils but it cannot be oxidized in anoxic environments. The nitrate in soils, because it cannot be held by ion exchange, usually percolates downwards, whereas the ammonia in sediments diffuses upwards to enter into oxic reactions at the sediment surface.

The rate of organic nitrogen mineralization has not been extensively studied probably because there is little that one can do to influence the rate, even if one wanted to do this, and because virtually no net accumulation occurs. On a yearly basis that which goes into the system is matched by an equal mineralization. This does not, however, mean that a knowledge of the process is unimportant in either system since time and rate of organic input may have a very marked effect on the rate and time of maximum mineralization, which in turn will influence the fate of the mineral nitrogen. Some of

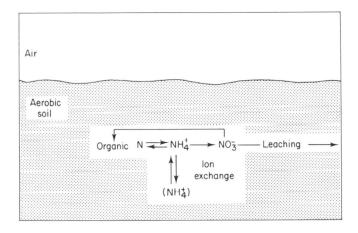

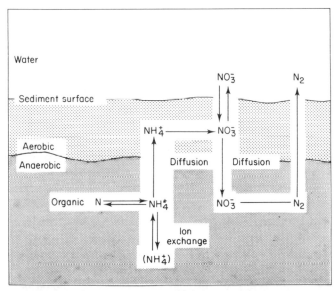

Fig. 29. The fate of nitrogen in a soil (top) and a sediment (bottom). In the aerobic soil the main loss of nitrogen occurs through leaching downwards. In the sediment losses will occur from diffusion of ammonia and nirate upwards but there is a greater possibility for denitrification in the anoxic sediment layers.

the factors affecting the fate of mineralized nitrogen are considered for the different systems, oxic soils and anoxic sediments.

5.1.1. Soil Mineralization

The process of mineralization in soil is less complex than in sediment, as the major portion of organic decomposition occurs at the soil surface. Some mixing by the soil fauna certainly occurs, carrying detritus to lower strata and some migration of bacterial cells would result in the same effect. Carbon is mainly oxidized by oxygen and the higher cell yields, as discussed in Chapter 3, result in approximately 50% of the organic or mineralized nitrogen being re-incorporated into microbial cells, if the material being decomposed has a C:N ratio similar to the cells. Individual systems vary with regard to the time of major organic input; presumably most root secretions occur during the growing season and leaf fall occurs during the autumn. The times for major decomposition would be at times of optimum rainfall and highest temperatures, which would vary with climatic conditions.

In agricultural soils, where the vegetation is removed by harvesting rather than allowing it to decompose in the soil, it is necessary to add nitrogen fertilizer to the soil to replace what has been removed. The high energy cost, and resulting high monetary cost, of making ammonia and nitrate fertilizers has made it more economical to consider again using animal wastes for this purpose. This is useful in another context, as the vast accumulations of cattle wastes from dairy units and feedlots, pose a dispersal problem. It becomes important to apply cattle manure in such a manner that its nitrogen becomes available in the soil when the agricultural crop needs it. This can obviously not be achieved with the precision of adding synthetic fertilizer, under controlled conditions, but the time and frequency of application is important. Pratt *et al.* (1976) made the following observations, based on a four-year trial with animal manures, applied to a sandy Californian loam soil: (1) The rate of irrigation was not important to crop yield (winter barley and summer sudan grass) but high application rates resulted in increased nitrate losses from the soil in all treatments. (2) Volatilization losses were trivial except when large excesses of manure were applied. (3) Annual application versus application every fourth year showed that the rate of nitrogen mineralization was much more constant for the more frequent applications, presumably resulting in a more constant supply of mineralized nitrogen. (4) All the nitrogen in dry dairy manure (1·6%N) was available in the first year of application whereas liquid feedlot manure (4·5% N) was 75% available.

In a rather different context (anaerobic and tropical), Yoneyama and Yoshida (1977) showed, using ^{15}N-labeled rice straw, that after the mineralization of the straw nitrogen, 25% was incorporated into plants after 130 days.

The rice straw decomposition resulted in a net uptake of fertilizer nitrogen, due to its low nitrogen content (0·5 % N). This bacterial uptake was maximum at 1 to 3 weeks and then no further uptake of ^{15}N-labeled ammonia from the fertilizer was observed. The conclusion may be drawn that when the bacteria have taken up sufficient mineral nitrogen for their growth requirements, this nitrogen remains, if not within the bacteria, at least within the straw residue, supplying further bacterial nitrogen growth requirements. The total nitrogen content of the residue remains the same but the percentage nitrogen increases.

In agriculture it is often useful to know how much nitrogen may be mineralized from the organic compounds in the soil, so that the fertilizer demand may be determined. The "mineralization potential" is usually measured by incubating the soil under standard conditions, and measuring the mineralized nitrogen. Stanford and Smith (1972) refer to some earlier experiments in this field and suggest a method based on periodic extractions of a soil sample, for up to 30 weeks, with $CaCl_2$ solution in a non-nitrogen nutrient medium. They found that the cumulative net N mineralized, from a large number of soil types, was linearly related to the square root of the incubation time. Mineralization potentials ranged from 20 to 200 ppm of airdried soils (5 to 40 % of the total nitrogen). There was little variation in the mineralization rate constant k which equalled $0·054 \pm 0·009$ week^{-1}, the time for half mineralization being $12·8 \pm 2·2$ week.

Nömmik (1976) found a good correlation between the nitrogen mineralized from acid forest soils during a 9-week incubation at 20°C and the carbon dioxide and ammonia released by a 2-hour digestion with a mixture of chromates and phosphoric acid, pH 5·95. Should such a method be applicable to a wide variety of soil types, it would be very useful for predicting nitrogen availability.

These data refer to net mineralization and give little information on the rates of immobilization (microbial uptake) and nitrification, and the effect on these processes of oxygen tension, redox potential, light and other factors which may influence the total balance. Some of these factors have been better explored in sediment systems.

5.1.2. Sediment Mineralization

As sediments are not used for agricultural purposes, the objectives in studying nitrogen mineralizations have been motivated by different objectives to those in soils. Experiments have largely been directed towards determining to what extent lake sediments can supply mineralized nitrogen, which may cause pollution problems. Also to what extent marine sediments can supply nitrogen to pass through various food chains and eventually maintain a commercially useful fish population. Lake and marine sediments differ in that, in the former, the oxidation of organic carbon is to a large extent coupled with methanogenesis; whereas, in the latter, sulfate reduction is relatively more important. The final result in both systems is the same, as a portion of the reduced carbon may be oxidized anoxically but reduced nitrogen (ammonia) is not oxidized under these conditions. Ammonia, therefore, accumulates, takes part in ion exchange reactions, but inevitably moves along a concentration gradient to the oxic surface layers (Fig. 29).

An example of such a gradient is illustrated in Fig. 30, showing the distribution of ammonia dissolved in interstitial water and held by ionic bonds to particles. In lakes similar concentration gradients occur with a similar distribution of ammonia between interstitial water and particles (Byrnes *et al.*, 1972). These authors demonstrated the capacity of anoxic sediments from Lake Mendota, to release ammonia to the overlying anoxic waters. An extension of this experimental approach confirmed that ammonia was liberated under anoxic conditions (Graetz *et al.*, 1973). The rate of ammonia liberation was relatively constant in laboratory experiments at 23°C, although the ammonia concentrations in the sediments increased significantly with time, as did the proportion of ammonia in the interstitial water. As the concentration of ammonia increased in the overlying water during the incubation, to a greater extent than it might have done in nature, the rate of transfer must be an underestimate. Some data for the four lakes are shown in Table XI. As one might anticipate, there is a correlation between the amount of ammonia mineralized and the rate of transfer from the sediments: Mendota deep sediment had the most rapid release of ammonia ($6 \cdot 02$ mmol NH_3

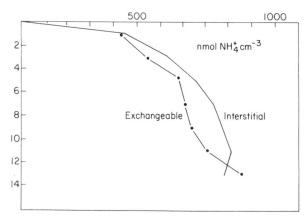

Fig. 30. The interstitial and the exchangeable ammonia profiles in a Limfjord sediment, June, 1977.

$m^{-2} day^{-1}$) and released the largest proportion of ammonia (32·9 %). In these experiments transfer was accomplished principally by the diffusion of ammonia, but in nature wave action, burrowing animals, bubbles from below etc. are likely to have a pronounced effect, in this instance in increasing the rate of transfer. In other instances, such perturbations may well have an opposite effect in oxygenating the sediment to greater depth, which would result in a conversion of ammonia to nitrate (making denitrification possible) and to an increased assimilative demand for mineralized nitrogen, by the more versatile and efficient aerobic flora.

A similar approach in measuring the rate of organic nitrogen mineralization is to fractionate sediment, incubate anaerobically at *in situ* temperatures and measure the net rate of ammonia production. If steady state conditions are assumed to exist in the sediment for some weeks, the prediction may be made that all of the excess ammonia will diffuse upwards and reach the sediment surface. An example of such an experiment is shown in Table XIII.

TABLE XI. Ammonia Transfer from Lake Sediment to Water.[a]

		NH_3 transfer Sediment to water ($mmol\, m^{-2}\, day^{-1}$)	NH_3 transfer in 48 days as % of total produced
Tomahawk	Soft water	1·45	10·6
Mendota	Hard water		
shallow sediment		2·78	11·5
deep sediment		6·02	32·9
Wingra	Hard water	3·62	19·9

[a] Data from Graetz *et al.* (1973).

The accumulated rate of ammonia production under 1 cm^2 of surface, down to 12 cm depth, is 0·438 μmol ammonia day^{-1} (or 4·38 mmol N m^{-2} day^{-1}). This value is of the same order as those in Table XI for lake sediments. The value for the upper 2 cm may be too high due to the death and decomposition of some aerobic microorganisms, the incubation being anaerobic. Not more than the upper 3 mm of sediment was oxic at this time of the year (Blackburn, 1979a).

The addition of ^{15}N-labeled ammonia to these laboratory incubations allows one to calculate, from the dilution of the label, the total rate of ammonia production (*d*) and the rate of ammonia incorporation (*i*), taking the increased pool size into account, and by measuring directly the net rate of ammonia production (*d* − *i*) (Blackburn, 1979b).

TABLE XII. Efficiency of C Incorporation Associated with Sulfate Reduction.[a]

Process	$\Delta G'_0$ (kcal mol^{-1})	ATP available for synthesis	C atoms dissimilated	C atoms assimilated
Fermentation: 100 hexose C → lactate + cells	− 52	2	88	12
SO_4^{2-} reduction: 88 lactate C → CO_2 + cells	− 38	2	69	19
Total: 100 hexose C → CO_2 + cells			69	31

[a] It is assumed that 1 mól ATP will form 10 g cells or 0·42 mol of organic C (Blackburn, 1979a).

Some average value are presented in Table XIII. There was a decrease in total ammonia production, going downwards in the sediment but there was still considerable activity at 10 to 12 cm depth. At this depth, however, there was little net ammonia production as ammonia incorporation was quite high; the latter showing surprisingly little variation with depth. Since the ratio of total production to incorporation varied with depth, this suggests that the C:N ratio of the detritus being decomposed might vary with depth. It is possible to calculate the composition of the detrital material being decomposed, as opposed to the general composition of the organic matter in the sediment, when the N:C ratio of the cells and the efficiency of incorporation are known (Section 3.3). The N:C molar ratio in the synthesized cells is taken to be 0·16 and the efficiency of carbon assimilation, associated with the combined processes of fermentation and sulfate reduction (the dominant processes in marine sediments) is derived as shown in Table XII. Carbon is incorporated at an efficiency of 31 %. These values are used as illustrated in Table XIII to calculate the substrate N:C ratio in the sediment, making the assumption that ammonia is the exclusive nitrogen source for the fermenting and sulfate-reducing bacteria. The measured N:C ratios are given for comparison. In both the calculated and measured values there is a decrease in the N:C ratio with depth, the effect being more pronounced in the calculated values. This indicates that the material which is most easily decomposed in the deeper sediment layers has a N:C ratio lower than the average.

The rate of sulfate reduction may be calculated from the rate of ammonia assimilation, again assuming that the fermenting and sulfate-reducing bacteria use ammonia exclusively as a source of nitrogen. The calculations

TABLE XIII. NH$_4^+$ Turnover, Calculated Rate of SO$_4^{2-}$ Reduction and Molar N:C Ratio of Substrate.[a]

Depth (cm)	Rate of NH$_4^+$ increase $(d-i)$ (nmol cm^{-3} day^{-1})	NH$_4^+$ production (d) (nmol cm^{-3} day^{-1})	NH$_4^+$ incorporation (i) (nmol cm^{-3} day^{-1})	N:C ratio calculated[b]	N:C ratio measured		SO$_4^{2-}$ reduction calculated[c] (nmol cm^{-3} day^{-1})
						$\frac{C}{N}$	
0– 2	74	131	57	0·11	0·11	9.09	396
2– 4	49	95	46	0·10	0·09	11.11	320
4– 6	40	78	38	0·10	0·10	10	264
6– 8	37	103	66	0·08	0·10	10	459
8–10	15	44	29	0·08	0·09	11.11	201
10–12	4	49	44	0·06	0·09	11.11	306
Total for 12 cm	438	1000	560				3992

[a] The rates are the average for four stations in the Limfjorden, August; data from Blackburn (1979a).
[b] (N:C in cells)/(N:C in substrate) × (efficiency of C incorporation) = (NH$_4^+$ incorporation)/(NH$_4^+$ total production). Efficiency of incorporation = 0·3; N:C ratio of cells = 0·16.
[c] 69 mol C oxidized ≡ 34·5 mol SO$_4^{2-}$ reduced ≡ 31 mol C incorporated ≡ 31 × 0·16 mol N incorporated (see Table XII). Thus 1 mol N incorporated ≡ 6·96 mol SO$_4^+$ reduced.

in Table XII indicated that $69 \, mol \, C$ were oxidized to accomplish the incorporation of $31 \, mol \, C$ into cells. The N:C ratio in cells is taken to be 0·16, giving $4·96 \, mol \, N$ incorporation. One mol $CH_2O - C$ oxidized is associated with the reduction of $0·5 \, mol \, SO_4^{2-} - S$, therefore, $1 \, mol \, NH^3 - N$ incorporated is equivalent to $6·96 \, mol \, SO_4^{2-} - S$ reduced. This relationship is used to calculate the rate of sulfate reduction in Limfjorden sediments at 2 cm intervals, from the rate of ammonia incorporation (Table XIII). The calculated values are higher than the measured values and do not show so marked a decrease with depth (Jørgensen, 1977a). The sum of the calculated rates down to 12 cm gives a value for sulfate reduction of $39·9 \, mmol \, S \, m^{-2} \, day^{-1}$, high in comparison with the average yearly rate of $9·5 \, mmol \, S \, m^{-2} \, day^{-1}$ (Jørgensen, 1977a).

This method of measuring rates of ammonia net production, total production and incorporation, thus allows one to make very useful deductions relating to the N:C composition of the material being decomposed and to correlate rates of incorporation with rates of carbon oxidation and sulfate reduction. Presumably a similar approach could be used in methane-producing systems to correlate ammonia turnover with methane production and carbon dissimlation.

An independent check may be made on the rates of the net ammonia production from a consideration of the depth profiles of interstitial ammonia concentrations. In the situations where ammonia is only produced in the deeper sediments and diffuses, without further net production or incorporation, to the surface, the gradient would be linear. A consideration of the profile in Fig. 30 shows deviations from linearity which are consistent with net ammonia production in the upper 10 cm of sediment. The magnitude of the change in gradient is proportional to the magnitude of production at any point and allows this rate to be calculated, knowing the diffusion coefficient of ammonia. There is, in general, quite good correlation between the measured rates and those calculated from gradients, indicating that both measurements give valid results.

These methods are best suited for upper sediments where biological activity is high. The measurement of rates of ammonia production is useful in defining the quantities of ammonia reaching the sediment surface, where further reactions may be studied. These reactions are so complex and difficult to measure, due to opposing oxygen and nutrient gradients, that it is very useful to know what the maximum rates from the ammonia flux might be. These ammonia production and incorporation rates, together with ammonia gradients, may be measured on a seasonal basis (varying light and temperature) thus giving information on the nitrogen status of the sediment over a yearly cycle. The annual turnover should almost equal the input but unfortunately the organic input is not always well defined since it may consist of pelagic, benthic and littoral material. By defining yearly ammonia mineralization rates, useful limits are set for the various inputs, taking sedimentation rates into account.

Berner (1974) has described how the diagenesis of ammonia may be quantified in deeper sediments where more constant conditions prevail.

5.2. NITRIFICATION

Nitrification is the process in which ammonia is oxidized, generally completely to nitrate. The importance of the process is that it produces an oxidized form of nitrogen which may participate in denitrification reactions, resulting in the loss of readily available nitrogen from the environment. There are, obviously, many other consequences of nitrification, which will be discussed, but they are of lesser importance when compared to this potential loss of nitrogen. One of the results of increased nitrate, whose importance is difficult to assess, is the possible selection of species which can readily assimilate nitrate. All plants and many bacteria have the ability to reductively assimilate nitrate, with some energy cost. Presumably if nitrification were never to occur, most plants and bacteria would have little difficulty in adjusting to using ammonia exclusively as a nitrogen source. It may, therefore, be deduced that in this context nitrification is relatively unimportant, operating only in having a minor selective influence on population composition and having little effect on total nitrogen cycling.

5.2.1. Microorganisms

There is little doubt that the lithotrophic autotrophic bacteria are mainly responsible for nitrification, but some heterotrophs, both fungi and bacteria, can oxidize ammonia (Alexander, 1965; Campbell and Lees, 1967; Painter, 1970). Fungi may be important in nitrification in acid conditions but the rate is slow and the eventual concentration of nitrate is low. Heterotrophic bacterial nitrification can lead to the transient accumulation of hydroxylamine in sewage, river and lake waters (Verstraete and Alexander, 1972) and some fungi can produce β-nitropropionic acid (Eylar and Schmidt, 1959).

The autotrophic oxidation of ammonia occurs in two steps, the first to nitrite by bacteria, physiologically represented by *Nitrosomonas* species:

$$NH_4^+ (-3) \rightarrow NH_2OH (-1) \rightarrow ?(+1) \rightarrow NO_2^- (+3)$$

The biochemistry of the process is poorly understood and is not important in the context of mineral cycling, as intermediates do not accumulate. The first enzyme involved in the oxidation of ammonia to hydroxylamine, may contain copper, as discussed by Campbell and Lees (1967). It is certainly inhibited by thiourea and by N-SERVE (2-chloro-6-(trichloromethyl)pyridine), the latter inhibition being relieved by copper ions (Campbell and Aleem, 1965). The reaction appears to involve the direct participation of oxygen:

$$NH_4^+ + 0.5O_2 \rightleftharpoons NH_2OH + H^+$$

Campbell and Lees (1967) consider that the reaction should be written as above, as hydroxylamine is not ionized at pH 7·0, and that the equilibrium lies to the left. There is, however, some doubt as to the free energy change in the reaction, possibly due to whether calculations consider the production of a proton or not. Painter (1970) reviews this point and quotes free energy changes of -0.7 and $+4.7$ kcal mol^{-1}. It seems certain, at least, that no ATP can be generated in the oxidation which Andersen (1965) considers to be endergonic. The overall free energy change in the oxidation to nitrite is $\Delta G'_o = -66$ kcal per mol of ammonia oxidized. The second nitrification step from nitrite to nitrate, performed by bacteria such as *Nitrobacter* species, involves a two-electron oxidation and is mediated through a cytochrome. No intermediates are involved:

$$NO_2^- + 0.5O_2 \rightarrow NO_3^- \qquad \Delta G'_o = -17.5 \text{ kcal.}$$

5.2.2. Growth

These autotrophic bacteria are very inefficient, relative to heterotrophs, in converting the energy available from their respective oxidations, into cellular material (Table XIV). *Nitrosomonas* generates 1 to 4% and *Nitrobacter* 3 to 10% of the cell material that might be expected from a heterotroph with the same amount of available energy. The reason for this is preumably that the autotrophs expend ATP and reducing power for the reduction of

TABLE XIV. Cell Yields of Nitrifying Bacteria.

	g cells/mol N oxidized[a]	Potential g cells/mol N oxidized[b]	Efficiency relative to heterotroph
Nitrosomonas	0·56–1·82	50	0·01–0·04
Nitrobacter	0·28–0·98	10	0·03–0·10

[a] Range of values quoted by Painter (1970).
[b] Calculated from the assumption that 50% of the $\Delta G'_o$, -66 kcal mol^{-1} for *Nitrosomonas* and -17.5 kcal mol^{-1} for *Nitrobacter*, is available for ATP production, that 7 kcal mol^{-1} are required for the synthesis of one mol of ATP and that 10 g cells are formed for each mol ATP, as in heterotrophic growth.

carbon dioxide; heterotrophs grow on already reduced carbon sources. The ecological implications of this low cell yield are that a small biomass of nitrifying bacteria turns over a very large amount of ammonia. Gundersen and Mountain (1973) give a value of one mol of bicarbonate incorporation for every 5 mol of ammonia oxidized to nitrate. Assuming that cells are 50% carbon, this would give 6 g cells per mol of ammonia oxidized ($\Delta G'_o =$ 83·5 kcal), or 1% of the efficiency of a heterotroph, based on the arguments in Table XIV. Associated with their poor growth yields are the slow growth rates of these bacteria. The range of growth rate constants (day^{-1}) reported for *Nitrosomonas* is 2·2 (Skinner and Walker, 1961) to 0·46 (Lees, 1952). Knowles *et al.*, (1956) give growth rate constants for *Nitrosomonas* and *Nitrobacter* at different temperatures whose values, in general, lie within these limits. These figures would correspond to mean generation times of 0·32 and 1·5 days respectively.

5.2.3.   Factors Affecting Growth of Nitrifiers

Substrate concentrations. Both oxidation stages are obligately dependent on oxygen but both can operate at low oxygen concentrations. The K_m values (Table XV) show quite a high affinity for oxygen, increasing with decreasing temperature (lower K_m). Nitrification can occur at oxygen concentrations as low as 6 μM. This is correlated with the occurrence of nitrification at Eh

TABLE XV. Substrate K_m Values for Nitrifying Bacteria.[a]

	Substrate	Temperature (°C)	K_m ($\mu mol\,l^{-1}$)
Nitrosomonas	O_2	30	16
		20	9
Nitrobacter	O_2	30	31
		18	8
Nitrosomonas	$NH_4^+ - H$	30	714
		25	250
		20	71
Nitrobacter	$NO_2^- - N$	32	600
		30	430
		28	360
		25	360

[a] Range of values quoted by Painter (1970).

values down to 210 mV at pH 7·5, in marine sediments (Billen, 1975). The affinity for oxygen of *Nitrobacter* is less than that of *Nitrosomonas* (this is not true for all the temperatures quoted in Table XV) which can result in the accumulation of nitrite under conditions of extreme oxygen limitation. Under normal circumstances, where oxygen is not limiting and the nitrification process is limited by the availability of ammonia (or carbon dioxide), nitrite does not accumulate.

The autotrophic nitrifying bacteria require three main substrates: carbon dioxide, oxygen and ammonia (or nitrite). It is unlikely that carbon dioxide is limiting in most environments but usually either oxygen or ammonia are in deficient supply. In well aerated sandy soils, ammonia is probably the limiting nutrient. Much of the ammonia is bound to negatively-charged soil particles where it may be inaccessible to *Nitrosomonas* although there is evidence that nitrification occurs at these sites (Lees and Quastel, 1946). In waterlogged soils and sediments, nitrification is limited by the rate of diffusion of oxygen into, and of ammonia through, the sediment and on local disturbances of the sediment surface. The depth of oxygen penetration and nitrification can vary from zero to 5 cm in the sediment, depending on the organic content; rich organic sediments, due to their high heterotrophic oxygen-demand do not allow oxygen to penetrate 5 mm below the surface (Vanderborght and Billen, 1975).

Nitrifying bacteria can survive for long periods in anoxic environments, even though they cannot grow (Painter, 1970). The entry of air into anoxic sediments and soils results in almost immediate nitrification, due to the presence there of nitrifying bacteria. This ability to survive in the absence of oxygen is particularly important in waste water treatment, where ammonia oxidation is desired after a period of anoxia, without resorting to inocula of nitrifying bacteria.

Temperature. Nitrosomonas grows best between 30 and 36°C with little growth below 5°C and a Q_{10} of 1·7 between 20 and 30°C (Buswell *et al.*, 1954). Nitrification was quite sensitive to temperature change in natural Thames water, the process having a Q_{10} of 2·7 (Knowles *et al.*, 1965) while the Q_{10} in activated sludge was 3·3 (Hopwood and Downing, 1965). *Nitrobacter* grows best between 34 and 35°C, with a range of 4 to 45°C (Deppe and Engel, 1960). It is believed, probably wrongly, that nitrification does not occur below 3 to 4°C. The nitrifying bacteria appear to be resistant to freezing, at least in natural environments, as ammonia oxidation is rapid when frozen soils thaw out. The rate of nitrite oxidation is more affected by low temperatures than is ammonia oxidation, which results in nitrite accumulation at low temperatures (Campbell and Lees, 1967).

Hydrogen ion activity. Nitrosomonas has a broad pH optimum, which may vary with strain, but usually lies between pH 6·0 and 9·0 (Engel and Alexander, 1958; Winogradsky and Winogradsky, 1933). Acid soils do not select for strains with acid pH optima (Meiklejohn, 1954). *Nitrobacter* strains have optima between pH 6·3 and 9·4 (Winogradsky and Winogradsky, 1933). Nitrite oxidation is reported to be inhibited more than ammonia oxidation at high pH values (this fact is not obvious from the pH optima) resulting in nitrite accumulation under alkaline conditions in excess of pH 8·5 (Campbell and Lees, 1967).

There are numerous literature references to nitrifying bacteria growing better on particle surfaces than in clear liquid media (Painter, 1970). It now seems probable that this is due to the pH buffering capacity of the particles. This is an important consideration as two protons are liberated in the oxidation of ammonium ion to nitrate, and acid conditions, which when produced in poorly buffered environments, can inhibit nitrification.

Carbon metabolism. The nitrifying bacteria are autotrophs; some aspects of their autotrophic metabolism are discussed in Section 1.2.1. It is probable that they are not sensitive to growth inhibition by small organic molecules, which is not surprising as they generally exist in a rich organic environment. Characteristically they adhere to particles, probably many of them organic, in both marine sediments (Vanderborght and Billen, 1975) and in soil (Gray *et al.*, 1968). It is indeed a characteristic of nitrifying bacteria that they are not found free in water but are found in river bottoms or in organic aggregates in estuaries (Tuffey *et al*, 1974). It is necessary for nitrifying bacteria to be in a region where ammonia is available, close to a zone of organic decomposition.

5.2.4. Quantification

There are surprisingly few quantitative determinations of the rate and extent of nitrification in natural ecosystems. Technically, it is difficult to measure since the product, nitrate, itself is involved in many biological reactions and therefore disappears. The alternative method for measuring nitrification is to calculate the rate of ammonia oxidation from the rate of ammonia production minus the rate of ammonia consumption, principally by incorporation into cells. In principle, this would appear to be a simple process but in practice it is not, due to the difficulties in using ^{15}N-tracer and because of the complex and opposing gradients of ammonia and oxygen that are normally found. An indirect, and potentially very useful method using ^{14}C-bicarbonate incorporation by normal and N-SERVE-inhibited sediments has been used to estimate nitrification rates in sediments (Vanderborght and Billen, 1975). These authors found nitrification only in sandy sediments and there only down to 5 cm. They did not determine rates at small intervals and therefore the effects of gradients were not obvious. They found nitrification rates of $1 - 8 \times 10^{-6}$ μmol ammonia oxidized cm^{-3} sec^{-1} (0.1–0.7 μmol cm^{-3} day^{-1}). These values when fitted to a mathematical model taking into account nitrification, denitrification, diffusion and sedimentation, very adequately described the nitrate profiles found in sandy sediments. These nitrification rates (0.4–3.5 μmol cm^{-2} day^{-1} for top 5 cm) are quite compatible with the rates at which ammonia is made available to the top 5 cm of sediment by diffusion from below and by *in situ* ammonia liberation (0.44 μmol cm^{-2} day^{-1} for less rich sediments; see Section 5.12).

A mathematical model for soil, taking into account position, time, rate constants for nitrification and denitrification, average solution velocities and distribution coefficient of ammonium ion between soil and soil solution and which describes the concentrations of ammonium ion, nitrite and nitrate at different times in steady state conditions, has been described by Cho (1971). The rate constant for ammonia oxidation, $k = 0.24$ day^{-1}, described by Cho may be compared with the values $k = 0.09$–0.69 day^{-1}, calculated from Vanderborght and Billen (1975), assuming a concentration for ammonia of 1 mM. Value of $k = 2$–5 day^{-1} for ammonia oxidation and slightly lower values for nitrite oxidation are reported for filter-grown populations of nitrifying bacteria (Srna and Baggaley, 1975).

Chen *et al.* (1972b) found no appreciable nitrification in acid, soft-water lake sediment, nor in hard-water lake sediments unless the latter were stirred to allow oxygenation. Under these conditions they measured nitrification rates (denitrification being ignored) of 0.17 and 0.46 μmol ammonia cm^{-3} day^{-1}, at 10°C and 25°C respectively. The ammonia concentration was approximately 1 mM, from which $k = 0.17$ and 0.46 day^{-1} may be calculated for the two temperatures, assuming that steady state conditions prevailed. These rates are summarized in Table XVI, giving in addition the turnover time. The k values are used in the calculations in Table XVII.

Nitrification in sediments is limited by the availability of oxygen, ammonia and a favorable temperature. These points are illustrated in the data in Table XVII, for a typical sediment in the Limfjord, Denmark. The higher temperature results in a greatly accelerated production of ammonia but because of the increased ammonia demand at the surface, the gradient is much steeper, resulting in a lower ammonia concentration close to the surface. Because of the increased biological activity, the depth of penetration of oxygen into the sediment is less at the higher temperature, as indicated by a steeper redox gradient. Nitrification is assumed only to occur at Eh values greater than 200 mV. The first order rate constants are chosen from the extremes in Table XVI and may bear no relation to those operating in the system. Using these constants, the calculated rates are very similar for the

TABLE XVI. Rates of Nitrification.

Free or interstitial NH$_4^+$ concentration (μmol cm^{-3})	Rate of NH$_4^+$ oxidation (μmol cm^{-3} day^{-1})	Rate constant (day^{-1})	Turnover time (days)
1·0	0·09–0·69	0·09–0·69	12–1[a]
—	—	0·24	4[b]
2·5	—	2–5	0·5–0·2[c]
1·0	0·17–0·46	0·17–0·46	6–2[d]

[a] Vanderborght and Billen (1975).   [c] Srna and Baggaley (1975).
[b] Cho (1971).   [d] Chen et al. (1972).

two temperatures and are in the range of 0·1–0·2 mmol N m^{-2} day^{-1}, or something like one third of the total ammonia production. The point to be emphasized is that even though both ammonia production and the rate constant for ammonia oxidation are greater at the higher temperature, these factors may be offset by the competition for oxygen and ammonia by other microorganisms.

5.2.5. Ecological Implications

The main ecological effects of nitrification, as discussed earlier, are that it can lead to denitrification. The probability of denitrification occurring in soils is increased by the lack of anion exchange capacity of soil. The tendency is, therefore, for nitrate to percolate down through the soil to increasingly oxygen-deficient zones, where denitrification can occur. Even in the absence

TABLE XVII. Calculated Nitrification Rates for Limfjord Sediments.[a]

	November	August
Temperature (°C)	5	18
Sediment depth for Eh = +200 mV (cm)	1·0	0·2
Average NH$_4^+$ concentration (μmol l^{-1})	0·2	0·1
k NH$_4^+$ oxidation[b] (day^{-1})	0·1	5·0
Total nitrification rate (mmol m^{-2} day^{-1})	0·2	0·1

[a] Data from Blackburn (unpublished).
[b] Values taken from Table XVI, covering the maximum range of rate constant values.

of denitrification, this leads to a loss of nitrogen from the upper soil in which the plant roots are located and to an eventual total loss of nitrate from the soil in run-off water. These drainage waters enter lakes and water systems which are often nitrogen-deficient, and thereby encourage unwanted algal blooms and eutrophication. The agricultural solution to this has been to use ammonia fertilizers and to inhibit nitrification with products such as N-SERVE.

The acidity produced in the oxidation of ammonia to nitrate may contribute to the lower pH of poorly buffered, non-calcareous soils, thus rendering them unsuitable for agricultural use. The extent to which this occurs is not well established. Low temperature, high pH and low oxygen tension may individually, or in combination, lead to an increase in nitrite by the inhibition of nitrite oxidation. Nitrite is a potent toxin and has adverse biological effects when its concentration builds up. The occurrence of nitrate, and potentially nitrite, in drinking water is thus a health hazard.

Nitrification is inhibited by a range of volatile sulfur compounds (carbon disulfide, dimethyl disulfide, methyl mercaptan, dimethyl sulfide and hydrogen sulfide) which may be produced from the decomposition of methionine and cysteine in soil (Bremner and Bundy, 1974). It is possible that these compounds may inhibit nitrification in natural ecosystems, but evidence is lacking as to the significance of the process.

5.3.   DISSIMILATIVE NITRATE REDUCTION

Dissimilative nitrate reduction (or the reduction of other nitrogen oxides, e.g., nitrite, nitric oxide or nitrous oxide) should be distinguished from assimilative reductions in which the reduced product (ammonia) is incorporated into cellular material. When the products of dissimilative reductions are the gaseous products, dinitrogen or nitrous oxide, the process is termed *denitrification*. Two types of dissimilative nitrate reductions may be recognised, (1) fermentative nitrate reduction and (2) respiratory nitrate reduction.

5.3.1.   Fermentative Nitrate Reduction

This term was used by Sato (1956) to describe a process in which no membrane-bound enzymes, cytochromes or electron transport phosphorylations were involved. Broda (1975b) emphasizes the difference between fermentative reductions and respiratory reductions, the latter involving membrane-bound enzymes, cytochromes, and electron transport phosphorylations. Nitrate may be considered as an "incidental" electron acceptor in the fermentative process, in much the same way that oxygen can be an "incidental" electron acceptor in other fermentations (Dolin, 1961). Both oxygen and nitrate act to reoxidize reduced NAD, thus sparing substrate which would have to be wasted, as for example in a lactic acid fermentation. This has similarities to the carbon dioxide reduction and hydrogen ion reductions discussed in Section 1.1.1, again involving a "fermentative" oxidation of reduced NAD. Ammonia is a typical, but possibly not obligatory, product of nitrate fermentation, as in *Clostridium perfringens* (Hasan and Hall, 1975). Ammonia can, however, also be produced as the end product of respiratory nitrate or nitrite reduction by *Aerobacter aerogines*, but no benefit accrues to the bacterium except in the reduction of nitrate to nitrite (Hadjipetrou and Stouthamer, 1965). Nitrate-respiring bacteria may produce ammonia under mildly anoxic conditions, possibly due to the sensitivity to oxygen of the low redox potential donors, involved in the production of dinitrogen and nitrous oxide (Campbell and Lees, 1967), but in general ammonia production from nitrate may be indicative of a fermentative pathway. There is little evidence that nitrate fermentation, as indicated by ammonia production is widely distributed in nature. Ammonia production has been demonstrated in air-dried soils (Stanford *et al.*, 1975a) and 30% of the aerobic isolates from a tropical mangrove sediment, produced ammonia (Nedwell, 1975). It has also been demonstrated in marine sediments (Koike and Hattori, 1978; Sørensen, 1978a).

Hydroxylamine is probably an intermediate in the fermentative dissimilative pathway, in the assimilative pathway and also in the pathway of ammonia oxidation. It does not appear to be liberated in normal respiratory dissimilation and possibly its presence may be indicative of a fermentative system. There is some evidence that the hydroxylamine that accumulated in the Japanese lake Kizaki-ko was derived from the reduction of nitrate (Tanaka, 1953). More recently hydroxylamine has been found in an Ethiopian lake, possibly from the same origin (Baxter *et al.*, 1973).

5.3.2. Nitrate Respiration

Bacteria. These are all bacteria which possess cytochromes. Some of the bacteria are strict anaerobes, viz. *Veillonella alcalescens* and *Selenomonas ruminatum* (de Vries *et al.*, 1974), yielding nitrite as an end product. Most species are, however, aerobic respirers which can anoxically use nitrate, instead of oxygen, as a terminal electron acceptor. Sometimes the reduction is only to nitrite, but often it proceeds to dinitrogen and is then termed denitrification. *Propionibacterium pentosaceum* is the only anaerobe known to be a true denitrifier (Van Gent-Ruijters *et al.*, 1975). Payne (1973) gives a comprehensive review of the variety of species involved in dissimilative respiratory reduction and the products of the reduction.

Regulation of denitrification. The key enzyme is nitrate reductase, typically a membrane-bound molybdo-iron sulfide protein, which reduces nitrate to nitrite, using a *b*-type cytochrome as an electron donor (Fig. 31). The synthesis of this enzyme and the cytochromes associated with denitrification, is depressed when oxygen tension is low. The process of nitrate reduction shows variability in oxygen sensitivity in different bacterial species but can occur at quite high oxygen tensions. In the presence of nitrate, the reduction of nitrite does not occur, due to partial repression of nitrite reductase. Nitrite accumulates until the nitrate has almost completely disappeared and similarly nitric oxide may accumulate until most of the nitrite is exhausted. The evidence that nitric oxide is an obligatory intermediate in the pathway

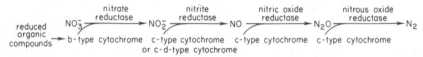

Fig. 31. The pathway of electron flow in a denitrifying system. The electrons from a reduced substrate flow through successive cytochromes, in the process reducing the oxides of nitrogen, mediated by a series of enzymes.

is not convincing, at least for *Pseudomonas aeruginosa* (St. John and Hollocher, 1977). Nitrous oxide is occasionally an end product of denitrification by species of the genera *Corynebacterium* and *Pseudomonas*, but usually it does not accumulate to a significant extent. It is, however, almost certainly an obligatory intermediate in the pathway leading to the production of dinitrogen. Evidence for this is seen in the ^{15}N-labeling pattern of N_2O and N_2 derived from NO_2^- (St. John and Hollocher, 1977) and the inhibition, by acetylene, of dinitrogen production by *Pseudomonas perfectomarinus* with the stoichiometric accumulation of nitrous oxide (Balderston *et al.*, 1976; Sørensen, 1978b). The less convincing evidence against the obligatory participation of nitrous oxide has recently been reviewed (Focht, 1974; Delwiche and Bryan, 1976).

Some bacteria are unable to reduce nitrate but can use nitrite (Vangnai and Klein, 1974). Many bacteria are able to reduce nitric and nitrous oxide even though they are unable to couple the reductions to energy-yielding reactions.

Energy yield. The free energy change ($\Delta G_o'$) for the oxidation of glucose by oxygen and nitrate are -686 and $-649\,\text{kcal mol}^{-1}$ respectively. Comparatively high growth yields would, therefore, be expected from growth on nitrate as the terminal oxidant. These yields are not always realized, particularly if nitrate has also to be assimilated (Sykes, 1975)· The growth efficiency depends on the extent to which phosphorylations can be coupled with electron transport. It has been suggested, as the nitrite reductase of *Pseudomonas denitrificans* is a soluble enzyme, that it was unlikely to be coupled with ATP generation (Yamanaka, 1964). The growth yields (g cells/electron equivalent) on nitrate, nitrite and nitrous oxide were 5·7, 5·6 and 4·4 indicating that phosphorylations were coupled to *each* reduction step (Koike and Hattori, 1975). This contrasts with the uncoupled reduction of nitrite to ammonia, by *Aerobacter aerogines* (Hadjipetrou and Stouthamer, 1965).

Factors affecting denitrification. The reduction of nitrate and other oxides of nitrogen demands a biological reductant. This is usually a reduced carbon compound as most of the bacteria performing the reductions are heterotrophs. The lithotrophic autotroph *Thiobacillus denitrificans* is also able to utilize reduced sulfur compounds (sulfide and thiosulfate) to reduce nitrate. Denitrification is thus limited to environments which contain reduced carbon or sulfur compounds. The effect of organic content and oxygen tension on rates of denitrification has been studied extensively in soils, sediments and waters (see Delwiche and Bryan, 1976). Not surprisingly, increased rates of denitrification are correlated with increased availability of respirable carbon and with decreased oxygen tension. The addition of sulfur to soils also increases denitrification (Mann *et al.*, 1972). Most of these rates have been measured using high concentrations of nitrate, 3·4 to 19·3 μmol nitrate per gram dried soil (Wijler and Delwiche, 1954), 1·8 mM nitrate to freshwater above sediments (van Kessel, 1977) and 0·16 mM ^{15}N-nitrate in Pacific waters (Goering and Dugdale, 1966a). Usually these concentrations are considerably in excess of the natural concentrations, the assumption having been made that denitrification rates are not affected by nitrate concentration (zero order kinetics). The reason for using these high concentrations of nitrate is that it has been extremely difficult to measure dinitrogen production in an atmosphere of 80% N_2, even when labeled with N-15, unless large quantities of nitrate were added. Unfortunately zero order kinetics probably do not normally prevail as is illustrated in Table XVIII, where some half-saturation constants (K_m) are given for systems demonstrating first order kinetics. Further examples are quoted by Yoshinari *et al.* (1977). It is likely that most denitrifying systems are operating under conditions of nitrate limitation, the enzymes being unsaturated, and the addition of extra nitrate will give erroneously high values. The solution to the problem may be to use ^{13}N-nitrate, in tracer amounts (Gersberg *et al.*, 1976), to increase the sensitivity of ^{15}N-detection or to construct model systems where curve fitting allows suitable rate constants to be estimated if other parameters are known (Vanderborght and Billen, 1975). An alternative, and very promising, system is the inhibition of dinitrogen production by acetylene, nitrous oxide being

TABLE XVIII. Half Saturation Constants (K_m) for Various
Denitrification Systems.

	K_m (mmol l^{-1})	
Sandy soil	12	Bowman and Focht (1974)
Estuarine Sediments		
Low organic	0·18	
High organic	0·6	Nedwell (1975)
Sewage	0·006	Moore and Schroeder (1971)
Freshwater sediments		
Low organic	4	
High organic	21	van Kessel (1977)[a]

[a] Approximate values based on Fig. 8 in van Kessel (1977).

released instead (Federova et al., 1973; Balderston et al., 1976; Yoshinara et al., 1977; Sørensen, 1978b). The method lends itself to field investigations, using undisturbed sediments and soils, the addition of acetylene being the only disturbance.

An in situ experimental procedure using a chamber located on top of the sediment appears to be a very useful approach in obtaining denitrification rates from the production of $^{15}N_2$ from $^{15}NO_3^-$ (Tirén et al., 1976). It has the disadvantage that it does not measure the amount of nitrogen produced from nitrate, generated within the sediment.

Denitrification has also been measured by N_2O reduction (Garcia, 1974, 1975), the presence of N_2O (Flühler et al., 1976) and argon to nitrogen ratios (Barnes et al., 1975). Rates of denitrification may also be deduced from the nitrate gradients in oceanic sediments (Bender et al., 1977) and in oceanic waters (Codisponti and Richards, 1976). Isotopic discrimination occurs during denitrification; $^{15}N/^{14}N$ ratios in N_2 have been used to measure denitrification (Wada et al., 1975)[†]

In sediments where nitrate production is limited to a narrow zone of 1 to 50 mm, depending on porosity, etc., denitrification is also limited, if it is dependent on in situ nitrate production. In the situation considered in Table XVII the estimated nitrification was only 0·1 to 0·2 mmol N m^{-2} day^{-1}, limiting denitrification to something less than this rate. Some idea of the actual rate may be obtained by using the first order rate constants for denitrification (k_d) of Vanderborght and Billen (1975), 0·17 to 3·46 day^{-1}. The nitrate concentration in Limfjord sediments typically varied from 5 to 20

[†] For most recent work on ^{15}N for denitrification studies see Koike, I. and Hattori, A. (1978), Appl. Environ. Microbiol. 35, 853–857 and Oren, A. and Blackburn, T. H. (1979), Appl. Environ. Microbiol 37, 174–176.

$nmol\,N\,cm^{-3}$ in the upper 1 cm, which would give a possible range for denitrification rates from 0·001 to $0·48\,mmol\,N\,m^{-2}\,day^{-1}$. The comparable denitrification rates, calculated from the data of Vanderborght and Billen (1975), taking an average of $50\,nmol$ of nitrate-$N\,cm^{-3}$ to a depth of 3 cm in muddy sediments, are 0·26 to $5·19\,mmol\,N\,m^{-2}\,day^{-1}$, a reflection of the greater activity of the Ostend Sluice Dock. Other values for denitrification are $0·003\,mmol\,m^{-2}\,day^{-1}$ for deep Atlantic sediments (Bender et al., 1977) and $\sim 4\,mmol\,m^{-2}\,day^{-1}$ for Danish lakes, on a yearly average (Andersen, 1977). A comprehensive review of denitrification rates, measured in sediments by a variety of methods, is given by Kamp-Nielsen and Andersen (1977) and by Brezonik (1973). The rates quoted range from 0·2 to $3·6\,mmol\,N\,m^{-2}$ day^{-1} for lakes to $72\,mmol\,N\,m^{-2}\,day^{-1}$ for a tropical estuary.

The rates for the Limfjord sediments, with a maximum denitrification rate of $0·48\,mmol\,N\,m^{-2}\,day^{-1}$ are of the same magnitude as the calculated rates for nitrate production, 0·1 to $0·2\,mmol\,N\,m^{-2}\,day^{-1}$, but neither figure should be taken too seriously as so many assumptions are made in their calculation. The seasonal variations (due to temperature and organic input) in the rate of ammonia production, nitrification and denitrification rate constants produce complex and changing gradients. It is not possible to calculate even ranges of possible rates without taking these factors into account.

With regard to the importance of measuring denitrification rates without increasing the nitrate concentration above the natural levels, it may be pointed out that when the rates were measured in Limfjord sediments using 0·5–3 mM ^{15}N-nitrate, the maximum rate of denitrification was $30\,mmol\,N$ $m^{-2}\,day^{-1}$. This is excessively high even though the nitrate concentrations added were modest in comparison to those used in many denitrification experiments. Using a K_m value at 0·6 mM (Nedwell, 1975) and an average nitrate concentration of $3\,\mu M$, the estimated rate of denitrification would be $0·14\,mmol\,N\,m^{-2}\,day^{-1}$.

A pattern that may be true for many systems would involve a winter rate of nitrification relatively higher than that of denitrification, resulting in an export of nitrate to the overlying water, this nitrate potentially diffusing back to the sediment in the summer, to be denitrified. The extent to which this occurs will depend on the extent to which the nitrate is transported away from the sediment and mixed with nitrate-deficient waters. In situations where the nitrate concentration in the overlying water is high, e.g., 5–60 μM, in a sewage polluted tropical estuary (Nedwell, 1975) or $100\,\mu M$ in the enclosed Ostend Sluice Dock (Vanderborght and Billen, 1975) extensive denitrification of the water-nitrate can occur. Denitrification will be greatly increased in sediments whose surfaces are disturbed by faunal or other agents (Vanderborght et al., 1977).

Denitrification occurs only at reduced oxygen tensions, mostly because the synthesis of the dissimilative enzymes is repressed by oxygen. Presumably under conditions of oscillating oxygen concentrations, such as might be induced by diurnal variations in hydrogen sulfide concentrations in sediments or by algal oxygen production, presynthesized enzymes in the reductive pathway could operate at oxygen concentrations that would have inhibited their synthesis. In natural systems such as soils and sediments, where denitrification mostly occurs, there is a requirement for oxygen to oxidize ammonia to oxides of nitrogen, before denitrification can occur. There is no doubt that complex interrelationships exist in anoxic sediments, as illustrated in Fig. 24. Ammonia which is produced in the anoxic depths of the sediment diffuses along a concentration gradient to the oxic interface where it is oxidized by nitrifying bacteria. The nitrogen oxides, mostly as nitrate, diffuse into the overlying water or downwards into an anoxic region, where they will be denitrified. The interesting situation thus arises in which oxygen inhibits denitrification but denitrification cannot occur without the participation of oxygen. The dependence of soil denitrification on oxygen has been nicely shown for model systems (Patrick and Reddy, 1967; Reddy and Patrick, 1976). In soils the presence of oxygen is related to how saturated they are with water. Oxygen can penetrate into waterlogged soils only by diffusion, a much slower process than by circulation through air spaces in the soil. Alternate flooding and drying of soils, sands or sediments is likely to encourage denitrification.

In general, denitrification goes most rapidly slightly above neutrality but can occur in the pH range 5·8–9·2 (Delwiche and Bryan, 1976). The most striking effect of pH is on the gaseous products formed, as discussed in Section 5.3.3.

Temperature has a marked effect on denitrification rates in soil (Stanford et al., 1975b). The Q_{10} (temperature coefficient) of denitrification in a variety of soils was 2, between 15 and 35°C. The rate of denitrification did not increase up to 45°C, and above that temperature there was a decrease in rate. There was a sharp fall in denitrification below 11°C with almost no activity below 5°C. Quite high rates of denitrification have been observed in sewage sludge at 5°C (Dawson and Murphy, 1972). A recent review by Focht and Verstraete (1977) contains further information on nitrification and denitrification.

5.3.3. Nitrous Oxide

The very imprecise measurements of denitrification in natural systems (and in the laboratory) have made it impossible to estimate directly the rates of terrestrial and oceanic dinitrogen production. Instead, indirect assessments are made on the basis that denitrification ($N_2 + N_2O$) exactly balances nitrogen-fixation. The calculated fluxes for terrestrial systems of N_2O, 16–69: N_2, 91–92, and for oceanic N_2O, 20–60: N_2, 5–99 million tons N year^{-1} (Söderlund and Svensson, 1976) give a much higher ratio of $N_2O:N_2$ than are normally found experimentally (CAST, 1976). The ratio $N_2O:N_2$ is very sensitive to pH, little nitrous oxide being produced at pH 7·0 but almost 100% at pH 5·0 (Focht, 1974). The oxygen tension also has a marked effect, the proportion of nitrous oxide rising with increasing oxygen (Focht, 1974). The inference may be that in natural ecosystems much denitrification may occur in partially oxic zones; this would be quite compatible with the previously discussed necessity for oxygen participation in ammonia oxidation, prior to denitrification.

The soil and oceanic waters may act as a sink for nitrous oxide removal from the atmosphere. It can be used in reductive assimilations or be dissimilated by the denitrifiers, if suitable conditions of oxygen tension, pH and reductant are available.

5.3.4. Other Biological Denitrifications

We have suggested (Chapter 2) that a complete nitrogen cycle could not occur on earth before the advent of oxygen and that denitrification is dependent on the presence of nitrogen oxides. The implication is that dinitrogen is not produced in totally anoxic environments into which neither oxygen nor nitrate can penetrate. Evidence from $N_2:Ar$ ratios indicates that dinitrogen is as conservative as argon, suggesting that dinitrogen is neither produced nor consumed in anoxic environments. There is, however, the interesting possibility that dinitrogen could be produced as follows:

$$NH_4^+ \rightarrow 0·5N_2 + 1·5H_2 + H^+, \qquad \Delta G'_0 = +9·5 \, \text{kcal (Broda, 1975b)}$$

The reaction would not appear to be particularly favorable but under conditions of active hydrogen removal, it might be a possibility. There are further possibilities of dinitrogen formation from ammonia oxidation by anaerobic photosynthetic bacteria and by chemolithotrophic bacteria oxidizing ammonia with nitrate (Broda, 1977).

There are also possibilities for non-biological denitrifications and nitrate reductions. These are not relevant in the context of this book and are thought to be insignificant.

5.4. NITROGEN FIXATION

Two of the most striking aspects of dinitrogen-fixation are (1) how quantitatively insignificant it is in comparison to ammonia flux and (2) how significant it is in maintaining a balanced nitrogen cycle.

The first point is illustrated from a consideration of the fact that terrestrial dinitrogen-fixation is only 194 million tons N year^{-1} compared with 8300 million tons N year^{-1} for ammonia mineralization (Rosswall, 1976). The major role played by symbiotic nitrogen fixation is discussed elsewhere (Section 8.2) and free-living bacteria are relatively unimportant. The mechanism of nitrogen fixation and the importance of the acetylene reduction assay in measuring this process has been discussed (Section 1.2.2). Hauck and Bremner (1976) review the literature relating the acetylene assay with $^{15}N_2$ reduction. They quote a range of factors from 2·4 to 25 for acetylene: dinitrogen, indicating that caution must be exercised in utilizing the theoretical ratio of 3, employed in most assays.

5.5. ASSIMILATIVE NITRATE REDUCTION

This is a process carried out by plants and some bacteria. It is not likely that nitrate is a source of cell nitrogen in anaerobic systems; the process is, therefore, only of importance in oxic systems. Quantitatively little is known as to what extent soil and marine bacteria use nitrate instead of ammonia, but the predominance of nitrate in well-aerated soils makes it likely that a significant portion of the 5800 million tons N year^{-1} (Roswall, 1976) incorporated by soil bacteria must be from nitrate. The reduction of nitrate to ammonia costs energy and ammonia is the preferred source of nitrogen. The presence of assimilative reductase (quite different from the dissimilative reductase) in bacteria may contribute to dissimilative nitrate reduction. This could happen in a situation where the dissimilative enzyme production was repressed by oxygen but a population capable of using the nitrite, generated by the assimilative enzyme, was present.

In general, assimilative nitrate reduction, whether in phototrophs or heterotrophs, results in the production of a reduced product (NH_3) which, on release during decomposition, can act as an energy source for nitrifying bacteria. In the case of the phototrophic plants, radiant energy is used, directly or indirectly, for nitrate reduction; the ammonia is thus analogous to reduced carbon as a carrier of solar energy, although quantitatively it is much less significant than reduced carbon compounds.

CHAPTER 6

The Sulfur Cycle

The sulfur cycle involves 8 electron oxidation/reduction reactions between sulfate ($+6$), the most oxidized form, and the most reduced form, sulfide (-2), the intermediate states being of lesser importance (Fig. 32). The cycle is principally driven by being linked to the oxidation of reduced carbon compounds (the products of plant photosynthesis) but it can also be fuelled by reduced carbon products from anoxic bacterial photosynthetic reactions, or by reduced sulfur compounds in sulfur springs.

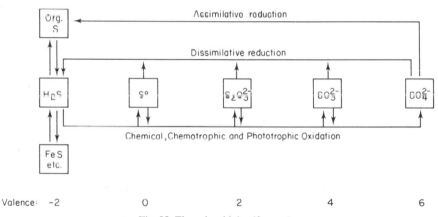

Fig. 32. The microbial sulfur cycle.

The most important aspect of the sulfur cycle involves dissimilative bacterial sulfate reduction and is discussed in Section 6.2. This concerns the mechanism whereby the major portion of the reducing power of reduced carbon compounds and hydrogen is transferred to hydrogen sulfide, which may then diffuse to an oxic site and participate in food chains. Sulfide thus acts as an energy carrier, allowing carbon mineralization to proceed. This type of process is restricted to anoxic environments which have a high sulfate content. In oxic environments the predominant sulfate reduction process is that of assimilation, principally by plants.

Both assimilative and dissimilative reductions yield sulfur at the -2 oxidation state, from which it is oxidized back to sulfate by microbiological and chemical processes. In this chapter we are mainly concerned with (1) dissimilative reductions and (2) sulfide oxidations, since they are uniquely microbial processes but it is useful to compare them quantitatively with the assimilative process.

6.1. ASSIMILATIVE SULFUR REDUCTION

Sulfate is assimilatively reduced to sulfide and reacts with serine to form cysteine which may then be converted to methionine. Both amino acids are normal constituents of protein and are the main sulfur-containing molecules in cells. The sulfur content of plants thus depends on the protein content and can vary considerably. Jørgensen (1977a) found 0·3% sulfur in eel grass (*Zostera marina*) while marine algae contain 0·3 to 3·3% (Goldhaber and Kaplan, 1974). Land plants vary considerably in their dry weight sulfur content but the N:S ratio will be approximately 30:1. The reductive assimilation of sulfate is thus much less important than the reductive assimilation of nitrate and consequently the oxidation of the protein-sulfide is less important than the oxidation of protein-nitrogen, when the plant residues are decomposed. In the Limfjord, Denmark, the decomposition of organic detritus accounts for less than 4% of the sulfide turnover (Jørgensen, 1977a). Similar estimates were made for the Black Sea (Deuser, 1970) and for anaerobic algal decomposition (Ramm and Bella, 1974).

6.2. DISSIMILATIVE SULFATE REDUCTION

Dissimilative sulfate reduction is dependent on the linked oxidation of reduced carbon compounds or hydrogen. The bacteria performing these oxidation/reduction reactions are described in Section 1.1 and their role in the carbon cycle is discussed in Section 4.1.4. The rates of sulfate reduction in the Limfjord on a yearly average, were 25 to 200 nmol SO_4^{2-} cm^{-3} day^{-1} (Jørgensen, 1977a), similar to values for a variety of marine sediments (Goldhaber and Kaplan, 1974). The rate of sulfate reduction clearly depends on the concentration of sulfate and on the reduced compound which is being oxidized. The availability of sulfate is dependent on the rate at which it can diffuse into the sediment from the overlying water. The availability of the reductant depends on more complex factors: the rate of sedimentation, the nature of the material being sedimented but principally on the rate at which small molecular weight molecules (hydrogen, methane, acetate, amino acids

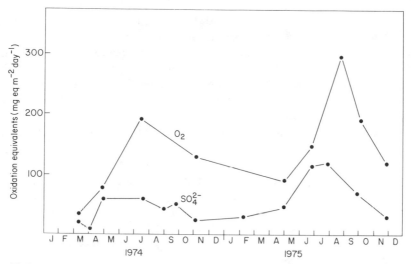

Fig. 33. Seasonal variation in the rate of sulfate reduction within the upper 10 cm of Limfjord sediment and in the dark oxygen uptake rate. The data points represent averages for several stations. (Redrawn from Jørgensen, 1977a.)

and sugars, etc.) are made available by hydrolytic and fermentative bacteria. These fermentative and hydrolytic processes linked to sulfate reduction gave temperature coefficients (Q_{10}) of 3·4 for Danish Limfjord sediment (Jørgensen, 1977a) and 3·9 for Wadden Sea sediment (Vosjan, 1975). Sulfate reduction thus shows a temperature response similar to the whole sediment activity as exemplified by the oxygen uptake in the Limfjord, which has a temperature coefficient of 3·2 (Jørgensen, 1977a). This is apparently in contradiction to the suggestion that dissimilative sulfate reduction is more sensitive to temperature than other heterotrophic processes in the sediment (Nedwell and Floodgate, 1972). The relatively equal dependence of both sulfate reduction and oxygen uptake on temperature is illustrated in Fig. 33. The oxygen uptake and sulfate reduction respond to seasonal temperature changes in an approximately equal manner. The difference between the two years was not due to differences in summer maximum temperatures but rather to differences in the organic input in the different years.

The process of dissimilative sulfate reduction has traditionally been associated with species of *Desulfovibrio* which can utilize only lactate, pyruvate, malate or hydrogen as reduced energy sources. Lactate has usually been the reduced carbon source included in isolation media and possibly species such as non-lactate utilizing *Desulfotomaculum acetoxidans* have been missed in counts of sulfate reducers (Sorokin, 1962; Bella *et al.*, 1972, Jørgensen, 1977a). In general, however, these authors have found a correlation between the presence of *Desulfovibrio* species and hydrogen sulfide. Jørgensen (1977a), however, did not find a good correlation between rates of sulfate reduction and *Desulfovibrio* counts—perhaps indicating that other (acetate- or methane-utilizing) bacteria may have been responsible for some of the sulfate reduction. He found no significant difference in *Desulfovibrio* numbers between sampling stations even though there was a considerable difference in the rates of sulfate reduction.†

In recent marine sediments, where sulfate is not limiting, the rate of sulfate reduction is greatest in the surface sediment. Jørgensen (1977a) showed that little activity occurred below 170 cm depth and that 65% of the total activity $(2 \cdot 06 \, \mu mol \, SO_4^{2-} \, cm^{-2} \, day^{-1})$ of a sediment column was located in the top 10 cm (Fig. 34).

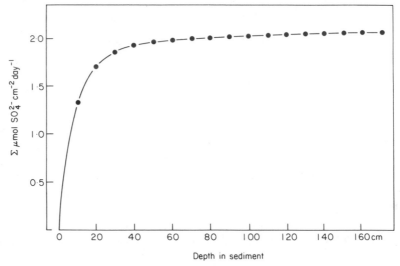

Fig. 34. Cumulative rates of sulfate reduction beneath 1 cm² of Limfjord sediment surface as a function of depth. (Redrawn from Jørgensen, 1977a.)

In sediments with active sulfate reduction, the sulfate concentration in the pore water decreases and the concentration of sulfide and thiosulfate increase. This is also true in stratified waters such as the Black Sea and the Cariaco Trench (Rozanov *et al.*, 1971; Tuttle and Jannash, 1973). Sulfate reduction also occurs in waterlogged soils; as with marine sediments, the quantity of sulfide production depends on the organic content of the soil (Ogata and Bower, 1965). The sulfide content can reach quite high levels; Sturgis (1936) found 29·9 ppm sulfide after three weeks of flooding in a clay loam, a con-

† For a comprehensive review of methods for quantifying sulfate reduction, see Jørgensen, B. B. (1979), *Geomicrobiol. J.* **1**, 11–64.

centration which can be toxic to plants. Sulfate reduction in soils is probably only quantitatively and economically significant in rice paddies but in some circumstances sulfide-corrosion of metals in soils and sewers can be a nuisance (see Section 7.1.3). Sulfate reduction in soils need not be in a completely anoxic zone and may occur in microniches (Wakao and Furosaka, 1976) as discussed in relation to marine sediments.

Sulfate reduction occurs in lake sediments (Cappenberg, 1975) but is of much less significance than in marine sediments due to the much lower sulfate content of fresh waters. The high sulfate content of stratified meromictic lakes leads, however, to a high rate of sulfate reduction in these systems.

6.2.1. Sulfide Ore Formation

Part of the hydrogen sulfide diffuses up to the oxic zone and part combines with iron to form iron sulfides. Jørgensen (1977a) calculated that 90% of the sulfide in Limfjord sediments is finally oxidized at the sediment surface, while about 10% forms iron sulfides (Fig. 35). Iron sulfides are also formed in waterlogged soils if sufficient iron is present (Connell and Patrick, 1969).

The process of pyrite formation is summarized in Fig. 36. Sulfide reacts with iron oxide, at neutral pH values, to form mackinawite, $FeS_{0.9}$:

$$2FeO.OH + 3HS^- \rightarrow 2FeS_{mack.} + S° + 3OH^- + H_2O.$$

The mackinawite is probably converted to greigite, although its formation

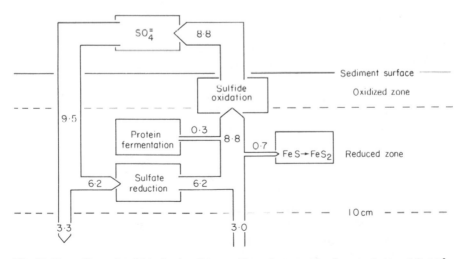

Fig. 35. The sulfur cycle of Limfjord sediments. The values are transfer rates in mmol S m^{-2} day^{-1} calculated as averages over two years for several stations. (Redrawn from Jørgensen, 1977a.)

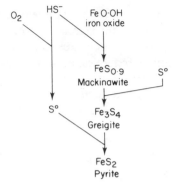

Fig. 36. The chemical processes leading to pyrite formation.

cannot be demonstrated in the laboratory. The reaction would proceed as follows:

$$3FeS + S° \rightarrow Fe_3S_4.$$

The greigite would react with further elemental sulfur to form pyrite:

$$Fe_3S_4 + 2S° \rightarrow 3FeS_2.$$

The sulfide in mackinawite and greigite may be released by mineral acids and is termed "acid-volatile sulfide". The sulfide in pyrite is not displaced in this way. At $< pH7\cdot0$ and at low sulfide concentration it is possible that pyrite may be formed directly without the formation of the acid-volatile sulfide intermediates.† The most readily available source of iron is the adsorbed oxide coating on clay minerals, but the clay minerals themselves contain much lattice-bound iron also, which can be exposed and react with sulfide. In general, it has been assumed that iron is not limiting in sedimentary pyrite formation (Berner, 1970) but the relatively minor (10%) quantity of sulfide trapped in pyrite in the Limfjord sediments would indicate that this is not universally true (Jørgensen, 1977a). The most active site for pyrite formation is the site of most active sulfide production which is near the sediment surface.†

The accumulation of pyrite in sediments and sedimentary rocks due to the ready availability of iron and sulfide is understandable. The accumulation of sulfides of copper (chalcopyrite and bornite), zinc (sphalerite) and lead (galena) is less readily explained, yet these minerals are thought to have formed as a result of bacterial sulfide production. These minerals occur in association with pyrite in stratiform deposits up to 100 km wide and from

† Pyrite formation, previously believed to be a relatively slow process may, in fact, take place rapidly following seasonal variation in sulfide concentration [Howarth, R. W. (1979), *Science* **203**, 49–51].

millimeters to meters in depth. Zinc is more abundant than lead, but copper when present may be the most abundant. Typically these deposits occur in association with carboniferous shales and dolomite, often close to hydrothermal activity. The normal concentrations of Zn, Pb and Cu are 80, 20 and 57 ppm respectively, insufficient to account for the 1% concentration often found in ore deposits. They would require a concentration factor of 125, 500 and 176 respectively to achieve the 1% level (Goldhaber and Kaplan, 1974). The only known site where heavy metal sulfides are at present being laid down is in the Red Sea (Degens and Ross, 1969), the source of the metals probably being basalt or Miocene shales. The process of transport to the sites of sulfide production is unknown but may be along geothermal gradients.

An alternative mechanism of concentration was suggested by Ferguson et al. (1975). In a model system they demonstrated that Pb and Zn were precipitated from an aerobic brine in which they were growing a species of *Chlorococcus*. The Fe-organic-carbonate sediment contained Pb at 0.5% and Zn at 1.0%, representing a concentration factor of 200- to 300-fold. The authors speculate that during early diagenesis, the metals may be expelled from the unstable carbonate phases to be converted into sulfides, possibly in veins.

Hallberg (1972) suggested that the organic molecules might chelate the metal and transport it from the site of sulfide precipitation to the sediment surface. This would represent a mechanism both for the spatial separation of different metals and also for their concentration at sediment surfaces in horizontal bands as oxides, which might in turn be converted to sulfides.

6.2.2. Hydrocarbon Oxidation

The capability of some sulfate-reducing bacteria to utilize methane as a sole energy source was discussed in Chapter 1. There is also evidence that strains of *Desulfovibrio desulfuricans* can slowly oxidize ^{14}C-methane, ethane and *n*-octadecane (Davis and Yarborough, 1966). This capacity of bacteria for linking sulfate reduction with hydrocarbon oxidation probably had very significant geological effects. The ubiquitous association of reduced sulfur compounds with petroleum deposits is due to the activity of these bacteria. Large quantities of reduced sulfur compounds accumulate only when sulfate is readily available, either by sulfate-rich waters percolating into the petroleum-bearing strata, or where the petroleum comes into contact with sulfate-rich deposits, such as $CaSO_4 \cdot 2H_2O$ (gypsum) in salt domes. In both situations, the hydrogen sulfide produced may be oxidized by thiobacilli (see Section 6.3) to elemental $S°$. When this process has been going on over a long period of time, it can result in large sulfur deposits. In the situation where gypsum is reduced, the sulfur is mixed with calcite ($CaCO_3$) whose $^{12}C:^{13}C$ ratio indicates that the carbonate was formed from the oxidation of an organic compound (viz. petroleum).

6.2.3. The Emission of Sulfur to the Atmosphere

Sulfide is readily oxidized by oxygen with thiosulfate as the principal product at pH > 8.5. Some sulfite is formed and is slowly oxidized to sulfate (Chen and Morris, 1972). There is thus little opportunity for hydrogen sulfide to escape to the atmosphere as its site of formation is usually overlaid by oxygen-containing waters. Where sulfide-rich coastal sediments or salt marshes are overlaid by shallow water, hydrogen sulfide can escape to the atmosphere in significant quantities. In two shallow coastal areas in Denmark the peak emission rates were 0·30 and 6·7 mmol $H_2S\,m^{-2}\,h^{-1}$ with diurnal averages of 0·063 and 1·58 mmol $H_2S\,m^{-2}\,h^{-1}$, respectively (Hansen et al., 1977). These emission rates represented 20% and over 100% respectively, of H_2S production (indicating that the latter rates were underestimates of the true rates of production). Very little sulfide could escape through oxic water 10 cm deep. The maximum emission occurred at night when photosynthetic oxidation of sulfide was not operating.

While the emission of H_2S can obviously be very significant in shallow sediments, as discussed above, it is unlikely to occur in the open oceans. In most marine environments dimethyl sulfide $(S(CH_3)_2)$ evolution is more important (Margoulis and Bandy, 1977). Again there is a diurnal pattern with maximum release just before sunrise. Dimethyl sulfide is produced by some marine algae and because of its resistance to oxidation and low solubility in water it would readily escape to the atmosphere. Oceanic water is saturated with dimethyl sulfide but its emission rate of 6 mg m^{-2} year^{-1} indicates that it alone could not account for the calculated emission rate from biological sources of 130 mg S m^{-2} year^{-1} (Margoulis and Bandy, 1977).

6.2.4. Isotopic Fractionation

There are four stable isotopes of sulfur; the most abundant are ^{32}S and ^{34}S. Some bacterial processes are able to discriminate between these forms, in general preferring the lighter ^{32}S isotope, which results in an enrichment of this isotope in the products of metabolism and an enrichment of ^{34}S in the unmetabolized sulfur substrate. As discussed in Section 2.2.3, these changes in isotope ratios are expressed as the enrichment of ^{34}S ($\delta^{34}S$‰) relative to a standard unmodified sulfur source. An increase in δ‰ value indicates an enrichment of ^{34}S. The changed ratios of $^{32}S:^{34}S$ in biologically processed

S-compounds, e.g., ores, can be a useful tool in establishing what process may have operated in their formation.

The type of fractionation and some values are presented in Fig. 37. There is only a small fractionation of sulfur isotopes in assimilative sulfate reduction by plants, animals and microorganisms. This is in contrast to the significant fractionation of carbon during reductive assimilation. The reduction of elemental $S°$, however, shows no isotope fractionation. This is evidence that $S°$ may be taken directly into the cell rather than being activated and thus fractionated, external to the cytoplasmic membrane. A small enrichment of ^{32}S occurs in the degradation of S-amino acids.

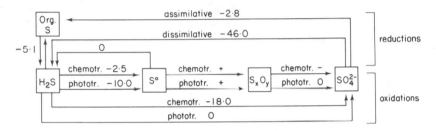

Fig. 37. Isotopic fractionation of sulfur by bacteria. The values represent $\Delta\delta^{34}S‰$. A negative value represents a preferential utilization of ^{32}S, a positive value a preferential utilization of ^{34}S, and a zero value indicates no preference. S_xO_y represents polysulfite. Chemotr = chemotrophic process; Phototr = phototrophic process; Org-S = organic sulfur. (Based on data from Gold-haber and Kaplan, 1974.)

There is a very significant selection of ^{32}S in dissimilative sulfate reduction, due to the large number of rate limiting steps that are involved, in many of which enrichment can occur.

Chemosynthetic oxidation of hydrogen sulfide by thiobacilli shows a slight preference for ^{32}S in the oxidation to $S°$ but the complete oxidation of H_2S to SO_4^{2-} has a $\Delta\delta^{34}S = -18‰$. In contrast, photosynthetic oxidation to $S°$ shows a significant preference for ^{32}S ($-10‰$) but there is no enrichment in the SO_4^{2-} product. The polythionates produced in both the chemosynthetic and photosynthetic oxidation contained an increase in the heavier isotope ^{34}S. This diversity in biological and abiological isotopic fractionation makes it difficult to attribute abundance in, e.g., a sulfur deposit, to a specific process.

6.3. CHEMOTROPHIC SULFUR OXIDATIONS

The pathways involved in the chemotrophic oxidation of reduced sulfur compounds, hydrogen sulfide (H_2S), elemental sulfur ($S°$) and thiosulfate

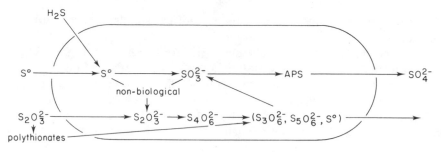

Fig. 38. Hypothetical pathways of oxidation of reduced sulfur compounds by a *Thiobacillus* cell.

$(S_2O_3^{2-})$ are poorly understood (Goldhaber and Kaplan, 1974). The main processes, as they occur in thiobacilli, are outlined in Fig. 38. Hydrogen sulfide can be oxidized non-enzymatically to $S°$:

$$2H_2S + O_2 \rightarrow 2S° + 2H_2O$$

and it is likely that $S°$ is on the main biological oxidation pathway. Elemental $S°$ itself may be oxidized by many thiobacilli and it is possibly first solubilized, without isotopic fractionation, and transported into the cell after reaction with glutathione or a protein with a reactive thiol group. Sulfur is probably oxidized by sulfite enzymatically to thiosulfate:

$$S° + SO_3^{2-} \rightarrow S_2O_3^{2-}.$$

The oxidation of thiosulfate yields complex products; tetrathionate $(S_4O_6^{2-})$, trithionate $(S_3O_6^{2-})$, pentathionate $(S_5O_6^{2-})$ and $S°$. The ratio and quantity of products depends on a variety of environmental factors; oxygen tension, pH, substrate concentration, etc. and some of these products can be produced abiotically outside the cell. It is likely that tetrathionate is the first oxidation product:

$$2S_2O_3^{2-} + H_2O + \tfrac{1}{2}O_2 \rightarrow S_4O_6^{2-} + 2OH^-.$$

The trithionate and pentathionate are formed from tetrathionate by reaction with sulfite and thiosulfate respectively:

$$S_4O_6^{2-} + SO_3^{2-} \rightleftharpoons S_3O_6^{2-} + S_2O_3^{2-}$$
$$S_4O_6^{2-} + S_2O_3^{2-} \rightleftharpoons S_5O_6^{2-} + SO_3^{2-}.$$

The contribution which chemotrophic bacteria make to the cycling of sulfur is difficult to assess as many sulfur compounds, notably H_2S are spontaneously oxidized by oxygen without bacterial mediation. There can be little doubt, however, that oxidations which occur slowly are speeded up by bacterial intervention as in the oxidation of sulfide ores by thiobacilli (Sokolova and Karavaiko, 1968). What is in doubt is the extent to which the bacteria benefit from these oxidations, especially with relation to growth yields on these substrates.

The chemotrophic oxidizers are a heterogeneous assortment of bacteria, the most important of which are the colorless sulfur bacteria, belonging to the families Thiobacteriaceae, Beggiatoaceae and Achromatiaceae (Kuenen, 1975). The genus *Thiobacillus* of the Thiobacteriaceae has been best studied and contains chemolithotrophic autotrophs deriving energy from the oxidation of reduced sulfur compounds and coupling this energy to the reduction of carbon dioxide. Thiobacilli can contribute very significantly to carbon dioxide fixation in stratified waters where hydrogen sulfide is produced in quantity, as in the Black Sea (Sorokin, 1965). At 160–200 m depth, at redox potentials of -20 to -40 mV, a rate of chemosynthesis of 0.3–0.5 mmol CO_2 m^{-3} day^{-1} was found. This was sufficient to regenerate the bacterial biomass once per day. While thiobacillus-like bacteria were present in this region, thiobacilli have not been isolated from the Black Sea and the rate of sulfide disappearance is the same with and without bacterial activity (Kuenen, 1975). Sorokin (1974) showed that thiosulfate oxidation may be more important than sulfide oxidation (which can be non-bacterial) and concluded that thiobacilli in the 150–180 m depth were probably responsible for thiosulfate oxidation and $^{14}CO_2$ uptake. Some of the $^{14}CO_2$ reduction may, however, have been due to methane-oxidizing bacteria (Sorokin, 1972).†

The quantitative contribution of thiobacilli to the oxidation of reduced sulfur in soils is not well documented except in very acid conditions where *T. ferrooxidans* and *T. thiooxidans* are important (Starkey, 1966). The contribution of the colorless sulfur bacteria is only known to be significant in environments which limit the growth of normal heterotrophic bacteria. Examples of such stressed environments are the acid soils quoted and hot acid sulfur springs, in which *Sulfolobus acidaldarus* occurs in almost pure culture (Mosser et al., 1973) In normal habitats, the presence of heterotrophic bacteria capable of oxidizing e.g., thiosulfate in soil (Trudinger, 1967) may complicate any assessment of the role of lithotrophic autotrophic bacteria. Similarly, an assessment based on $^{14}CO_2$ uptake may be complicated by heterotrophic CO_2 assimilation, which can be 3 to 6% of their cell carbon, and their numbers and growth rate may far exceed those of the lithotrophic autotrophs.

The obligate chemolithotrophic, autotrophic thiobacilli are not unique in being capable of chemotrophic sulfur oxidations. Kuenen (1975) lists the other groups as (1) mixotrophs (capable of heterotrophic/autotrophic

† In a recent detailed study of sulfur oxidation in a chemocline [Jørgensen, B. B., Kuenen, J. G., Cohen, Y. (1979), *Limnol. Oceanogr.* **24**, (in press)] thiobacilla are shown to catalyze the initial oxidation of H_2S.

growth), (2) chemolithotrophic heterotrophs, (3) heterotrophs which do not gain energy from the oxidation of sulfur compounds but may benefit from the oxidation, and (4) heterotrophs which gain nothing from the oxidation.

It is not obvious where bacteria such as *Thiothrix* and *Beggiatoa* fit into this scheme but it is not yet proved that they derive energy from hydrogen sulfide oxidation and may benefit by using the reaction to remove toxic hydrogen peroxide (Tredway and Burton, 1974). The large biomass of *Beggiatoa* species associated with some marine sediments (Jørgensen, 1977b) was discussed in relation to the carbon cycle (Section 4.2.3).

The problems of differentiation between bacterial and chemical oxidation of sulfur compounds makes it difficult to assess the importance of bacteria in sulfur cycling in general. Bacteria are certainly important in the oxidation of sulfide ores, elemental sulfur and probably thiosulfate. This, and the ecological disadvantages of acid production associated with the oxidation of mineral sulfides are discussed in Section 6.2.2. The acidity associated with the oxidation of elemental sulfur has been used to advantage in soil, by producing an acid environment inhibitory to *Streptomyces*, plant pathogens (Alexander, 1961). In addition, acid conditions may increase the availability of phosphate.

Oxygen is not the only oxidant for sulfur compounds. Nitrate can act as the terminal electron acceptor in the oxidation of hydrogen sulfide, sulfur, thiosulfate, sulfite, tetrathionate and dithionate by *Thiobacillus denitrificans*, in the absence, of oxygen at neutral pH values; for example:

$$5S_2O_3^{2-} + 8NO_3^- + H_2O \rightleftharpoons 10SO_4^{2-} + 4N_2 + 2H^+,$$

$$\Delta G_o' = -893 \, \text{kcal}.$$

Sulfide oxidation coupled to nitrate reduction might be expected to be an important process in some oceanic situations where sediment-produced sulfide might meet nitrate-rich, oxygen-deficient overlying waters. No estimate is available regarding the extent to which this is important on a global scale but in the Black Sea and Cariaco Trench it apparently does occur (Tuttle and Jannash, 1973). Thiobacilli-like bacteria were isolated from these waters which were facultatively autotrophic, oxidizing sulfide and thiosulfate under anoxic conditions using nitrate as oxidant. They did not produce acid conditions, indicating an incomplete oxidation and their ability to oxidize thiosulfate was limited, thus explaining the relatively high thiosulfate content at certain sites. There was evidence that these, or other, bacteria were capable of hydrogen sulfide oxidation at depths where both oxygen and nitrate were absent, suggesting the possibility that another oxidant was involved. It is unlikely that sulfate is involved as an electron acceptor as:

$$HS^- + SO_4^{2-} + H^+ \rightleftharpoons S_2O_3^{2-} + H_2O, \qquad \Delta G_o' = +4.5 \, \text{kcal}.$$

The reaction would not appear to be thermodynamically favorable. Possibly hydrogen sulfide oxidation is coupled to the reduction of an organic compound (Tuttle and Jannash, 1973).

6.4. PHOTOTROPHIC SULFUR OXIDATIONS

The types of bacteria involved with phototrophic sulfur oxidations are briefly discussed in Section 1.1.2 and the photoprocess outlined in Fig. 3. As discussed in Section 4.2.2, habitats in which anoxygenic photosynthesis is important are relatively limited in distribution and as a result, this process does not contribute significantly to either global carbon- or sulfur-cycling. In limited situations it can, however, be very important and is worth consideration in some detail.

6.4.1. Sulfureta

The process of photosynthetic sulfur oxidation is dependent on light and on a source of reduced sulfur compounds, supplied either by abiotic processes (sulfur springs, etc.) or by anoxic bacterial sulfur reductions. This type of habitat, dominated by bacterial sulfur oxidations has been given the name "sulfuretum" by Baas-Becking (1925). A general account of the natural history of sulfureta is found in Fenchel (1969).

The distribution of sulfur springs is very limited but some interesting bacteria such as the thermophile *Chloroflexis* are prevalent in hot springs in North America, Iceland and New Zealand. These bacteria are usually associated with cyanobacterial mats and grow photoheterotrophically but they are also capable of phototrophic sulfide oxidation (Castenholz, 1973).

The distribution of biogenic sulfureta is more widespread. They are found on muds and sands, particularly marine, where sulfide diffuses from below and where oxygen penetration of the surface is low. Sulfureta also develop in completely aquatic environments where mixing of the water never occurs, as in meromictic lakes, and where seasonal stratification can occur, as in holomictic lakes. In these aqueous environments the phototrophic bacteria are located in a narrow horizontal plate where light, hydrogen sulfide and minimal oxygen concentration are most favorable for their multiplication. They possess a range of pigments capable of capturing light mostly of wavelenths that pass through any oxygen-evolving photosynthetics which may overlie them. There is, however, a limit to light penetration and the bacterial photosynthetic plate rarely lies below 20 m. The photosynthetic bacteria are anaerobes and are thus limited to anoxic environments. They may also be inhibited by hydrogen sulfide if in too high a concentration. The limits for pure cultures are in the range of 0·4 to 8 mM hydrogen sulfide (Hansen and Van Gemerden, 1972), the green sulfur bacterium *Chlorobium* being most tolerant (4–8 mM), the purple sulfur bacteria *Chromatium, Thiocystis* and *Thiocapsa* less tolerant (0·8–4 mM) and the purple non-sulfur bacteria being the least tolerant (0·4–2 mM) (values quoted by Pfennig, 1975).

In general the main reduced sulfur compounds available are hydrogen sulfide and elemental sulfur. All these bacterial types can oxidize hydrogen sulfide usually with the intermediate production of intracellular S° (purple sulfur bacteria) extracellular S° (the purple sulfur bacterium *Ectothiorhodospira* and the green sulfur bacteria). This S°, or other exogenous S°, can then usually be oxidized further to sulfate but strains of *Chlorobium vibrioforme* have only a limited capacity to oxidize S° and the purple non-sulfur bacteria are completely incapable of S° oxidation (see references in Pfennig, 1975).

Because of the higher tolerance of the green sulfur bacteria to hydrogen sulfide, they are usually located in sulfureta below the purple sulfur bacteria.

6.4.2. Sulfur Turnover

In the relatively closed systems in which the sulfureta exist, there may not be a requirement for a large amount of total sulfur to be present in order to maintain a high population of phototrophic sulfide oxidizers. The reason for this is that the sulfate which is produced can be reduced again by anaerobic bacteria such as *Desulfovibrio* species. It is quite common to find both purple and green sulfur bacteria associated in a tight complex with sulfate reducers, the photosynthetics supplying reduced carbon to the sulfate reducers which in turn supply sulfate to the photosynthetics (Pfennig, 1975). Moreover, in a culture of such mixed species (*Chromatium* and *Desulfovibrio*) very small concentrations of sulfate $(0·015 \, \text{mmol} \, l^{-1})$ maintained high population numbers, the sulfate being recycled once every 15 min (Van Gemerden, 1967). Presumably larger quantities of sulfate would be required in natural systems in which exchange of sulfate and sulfide between the sites of photosynthesis and sulfate reduction would be limited by diffusion. The rates of sulfide oxidation can be very large and presumably some portion of this must be from recycled sulfate. As an example, in the Fayetteville Green Lake, New York (Culver and Brunskill, 1969) the quantity of sulfide oxidized to reduce $47·0 \, \text{mmol} \, C \, m^{-2} \, day^{-1}$ (average) would be 84 tons H_2S $year^{-1}$ for the whole lake area of $0·25 \, km^2$ (Pfennig, 1975). A major portion of this quantity must have been recycled within the year. In Fayetteville Green Lake, a meromictic water body, 83 % of the total primary productivity was due to photosynthetic sulfide-oxidizing bacteria at 18–20 m depth (Culver and Brunskill, 1969). These bacteria were preyed on by various zooplankton (*Daphnia, Orthocyclops, Filinia*) which were found just above the bacterial plate. The zooplankton in this type of environment can themselves be a source of food for fish.

Czeczuga (1968) measured rates of bacterial carbon fixation at 4 m depth, between 0.36 and 3.9 mmol C m^{-2} day^{-1} in Lake Wadolek, Poland. While these rates are relatively lower than in Fayetteville Green Lake, they represented a significant proportion of the primary productivity; 166.6% in June and 13.4% in July of the phytoplankton productivity. In Lake Popowaka Mala, Poland, the *Chlorobium* productivity reached 12.64 mmol C m^{-2} day^{-1}, corresponding to 25.28 mmol H$_2$S m^{-2} day^{-1} oxidized.

6.4.3. Competition for Sulfide

The colorless sulfur bacteria occupy a niche slightly different from that of the phototrophic sulfide oxidizers, as they require oxygen. The two types are, however, commonly found closely associated. In anoxic sediments which have been freshly mixed by natural turbulence or artificially, *Beggiatoa* are often the first to re-establish themselves close to, or on the sediment surface (Fenchel, 1969; Jørgensen and Fenchel, 1974). As the sulfide concentration increases and the sediment becomes more anoxic, the *Beggiatoa* species moves to the oxic surface and the photosynthetics are found immediately below them. Finally the *Beggiatoa* mat dies and the photosynthetics predominate. The two populations can exist synchronously in situations where the sulfide concentration is kept low and the oxygen tension at the sediment surface high, by the sulfide oxidizing action of the phototrophs. During the night the sulfide concentration increases and the oxygen decreases, forcing the *Beggiatoa* species to migrate upwards above the sediment surface (Hansen *et al.*, 1977). The combined action of the phototrophs and the colorless sulfur bacteria is to lower the sulfide concentration to a level no longer toxic to other forms of life and in the process generate biomass which may be used by higher forms of life.

CHAPTER 7

Other Elements

This chapter describes a heterogeneous collection of chemical reactions mediated by microorganisms in nature. While the cycling of carbon, sulfur, oxygen, and nitrogen to a large extent can be described in terms of energy yielding oxidation–reduction reactions and assimilatory reductions, the processes described in the following are of a more varied nature. Thus, in addition to dissimilatory reductions and oxidations they include various assimilatory processes, processes mediated indirectly through microbial activity, microbial detoxification of certain compounds and some processes which are incompletely understood.

7.1. THE MICROBIAL OXIDATION AND REDUCTION OF METALS

A number of processes involving the reduction or oxidation of various metals are catalyzed by bacteria in nature. The significance of this is not always clear. Traditionally a number of bacterially mediated oxidations of Fe^{2+} and Mn^{2+} ions have been considered as examples of chemolithotrophy. Many studies within the last decades have only established one of these processes, viz. that of thiobacilli oxidizing ferrous iron in acid environments, as a true example of chemolithotrophy. This process will, therefore, be discussed separately in the following section. The two following sections will then treat other microbially mediated oxidations or reductions of metals.

7.1.1. The Pyrite Oxidation Cycle

The oxidation of ferrous iron by oxygen is an exothermic reaction which can be described by:

$$O_2 + 4Fe^{2+} + 4H^+ \rightarrow 4Fe^{3+} + 2H_2O.$$

At pH values $> 5 \cdot 5$ the spontaneous oxidation takes place rapidly and the rate is strongly pH dependent; in fact:

$$-d[Fe^{2+}]/dt = k[Fe^{2+}] [OH^-]^2 pO_2,$$

where $k = 8 \times 10^{13}$ min^{-1} atm^{-1} mol^{-2} at 20°C. Thus the rate increases by a factor of 100 for each unit increase of pH. The process is driven by the hydration of Fe^{3+}:

$$Fe^{3+} + 3H_2O \rightarrow Fe(OH)_3(s) + 3H^+,$$

due to the very small solubility product of hydrous ferric oxide in neutral or alkaline environments. In such environments the oxidation of ferrous iron, as an energy yielding process for microorganisms, is therefore, questionable and also very difficult to study. At pH values of 2–3·5, however, the oxidation rate is rather independent of pH and is very slow if not catalyzed, having a half life of the order of 1000 days, at a pH value of 3 (Stumm and Morgan, 1970). The process will yield about 10·6 kcal per mole iron oxidized. This is utilized by *Thiobacillus ferrooxidans*, an organism which can also oxidize reduced sulfur compounds and elemental sulfur. It has been shown that the organism can grow autotrophically on Fe^{2+} and that it then assimilates CO_2 through the Calvin cycle (Maciag and Lundgren, 1964; Lundgren *et al.*, 1972; Belly and Brock, 1974).

Thiobacillus ferooxidans is found where pyrite (FeS_2) is exposed to atmospheric oxygen. In anaerobic sediments biologically produced HS^- will precipitate ferrous iron as FeS. This is then transformed abiologically at a low rate into FeS_2 (see Section 6.2.1). Pyrite will come in contact with atmospheric oxygen where such fossil or recent anaerobic sediments are exposed, e.g., in drainage channels and in lignite and coal mines which may contain considerable amounts of pyrite. In mines where sulfide ores are exploited, large amounts of pyrite may also come in contact with oxygen.

The initial process of the pyrite oxidation carried out by *T. ferrooxidans* is the oxidation of S^{2-}:

$$2FeS_2(s) + 2H_2O + 7O_2 \rightarrow 2Fe^{2+} + 4SO_4^{2-} + 4H^+. \tag{1}$$

The bacterium can then further oxidize the ferrous iron according to:

$$4Fe^{2+} + O_2 + 4H^+ \rightarrow 4Fe^{3+} + 2H_2O. \tag{2}$$

The Fe^{3+} formed (which will remain in solution at the low values of pH) is itself a strong oxidant which abiologically will oxidize pyrite according to:

$$FeS_2(s) + 14Fe^{3+} + 8H_2O \rightarrow 15Fe^{2+} + 2SO_4^{2-} + 16H^+, \tag{3}$$

yielding more ferrous iron which serves as a substrate for *T. ferrooxidans*. As the pH of the effluent water increases, the ferric iron formed is hydrated according to:

$$Fe^{3+} + 3H_2O \rightarrow Fe(OH)_3(s) + 3H^+, \qquad (4)$$

leading to the precipitation of hydrous ferric oxide.

The whole process (1)–(4) is called the *pyrite oxidation cycle*. The process can proceed abiologically but the rate is greatly increased (perhaps by a factor of the order of 10^6) through the activity of thiobacilli. The main contribution of these bacteria is the oxidation of ferrous iron which is the rate limiting step in the prevailing acid environment, whereas the oxidation of pyrite by ferric iron takes place rapidly without biological catalysis. The pyrite oxidation cycle is shown schematically in Fig. 39. Other thiobacilli (*T. thiooxidans*) which are active at low values of pH may also be involved in the pyrite oxidation cycle by oxidizing various reduced sulfur species and elemental sulfur appearing during the process, the role of this in the overall process, however, is less significant. The pyrite oxidation cycle is the most acid-producing, biologically mediated, oxidation process known. Thus the steps (1)–(4) lead to the production of 5 protons for each atom of iron which is oxidized.

In some acid mine waters it has been found that another bacterium assigned to the genus *Metallogenium* (see Section 7.1.2) plays a role in the oxidation of ferrous iron. This filamentous bacterium seems to be active in the pH range of 3·5–4·5 whereas *T. ferrooxidans* is active at pH $< 3·5$. Thus *Metallogenium* precedes *Thiobacillus* in a succession when pyrite oxidation is initiated. The physiological role of the iron oxidation for *Metallogenium* is unknown (Walsh and Mitchell, 1972).

The microbially catalyzed pyrite oxidation has negative as well as positive practical consequences. In the surroundings of coal, lignite, and sulfide ore mines, or where river beds have been moved or drainage channels made, the process may become an environmental nuisance (e.g., Lundgren *et al.*, 1972). Streams and rivers may become acid beyond the tolerance of the native freshwater fauna and the excessive precipitation of ferric hydroxides, which is also aesthetically unpleasant, may have similar effects (Fig. 40). When exploiting low-grade sulfide ores, however, the process may be very useful. Ferric iron will not only oxidize pyrite but also other metallic sulfides and UO_2. Some of these processes are shown in Table XIX. The bacterially produced Fe^{3+} therefore acts as a leaching agent, solubilizing various useful ores which are not used by thiobacilli as substrates. The significance of microbial ore leaching has been studied intensively and it is employed in many mines for recovering pyrite-containing, low-grade ores of zinc, copper and other metals (see Silverman and Ehrlich, 1964; Zajic, 1969 and references therein). It is important that the ore is granulated in order to increase the surface area and also that the mine water be kept oxygenated. The dis-

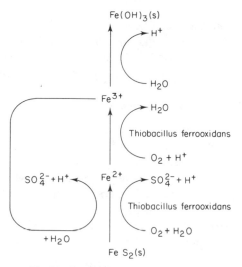

Fig. 39. The pyrite oxidation cycle.

solved metal ions (e.g., Cu^{2+}) may be recovered from the effluent mine water by passing it over scrap iron where the process:

$$Cu^{2+} + Fe^{\circ} \rightarrow Fe^{2+} + Cu^{\circ}$$

takes place.

The importance of *T. ferrooxidans* in these processes also depends on the

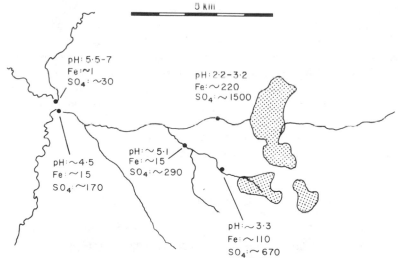

Fig. 40. The values of pH, suspended + dissolved iron and SO_4^{2-} concentrations (both in ppm) in streams draining abandoned lignite mines (shaded area) in Jutland, Denmark. (Data from Jacobsen, 1975.)

extraordinary tolerance to metallic ions shown by this bacterium. According to Tuovinen et al. (1971), the organism can tolerate concentrations of Zn^{2+}, Ni^{2+}, Cu^{2+}, Co^{2+}, Mn^{2+} and Al^{3+} up to $10 g l^{-1}$; Ag^+, UO_2^{2+}, AsO_2^-, SeO_2, and TeO_3^{2-} up to $50-100 mg l^{-1}$, and MoO_4^{2-} of up to $5 mg l^{-1}$.

7.1.2. The Reduction and Oxidation of Iron and Manganese

A very large and heterogeneous assemblage of bacteria are known to deposit $Fe(OH)_3$ or MnO_2 from dissolved iron (Fe^{2+}) and managanese (Mn^{2+}). Both of these reactions are exothermic, albeit with a small yield of free energy. Since Winogradsky (1888) suggested that these processes exemplify chemolithotrophy, the question has been studied by many authors. The study of the physiological significance of the processes has proved difficult. As already discussed in the previous section, the oxidation of ferrous iron, often presented as the oxidation of siderite:

$$4FeCO_3 + O_2 + 6H_2O \rightarrow 4Fe(OH)_3 + 4CO_2$$

takes place spontaneously in neutral environments where most of the bacteria discussed below occur. Manganese ions do not oxidize spontaneously in neutral solutions. However, the presence of MnO_2 in cultures poses difficulties with respect to the measurement of cell yield as discussed below. The best studied group within the "iron bacteria" (excepting *Thiobacillus ferrooxidans*) is constituted by the "sheathed bacteria" or Chlamydobacteria. Their taxonomy, physiology and ecology have been reviewed by Pringsheim (1949), Mulder (1964) and most recently by Dondero (1975).

The bacteria are rods which are found surrounded by filamentous sheaths; the bacteria are usually found in single, end to end rows but the filaments may show false branching. Aggregates of filaments may assume macroscopic dimensions. The bacteria have been assigned to a number of genera (*Sphaerotilus, Leptothrix, Cladothrix, Crenothrix*) but these genera have mainly been created on the basis of morphological features of bacteria in natural samples, and many of them have not been grown in pure cultures. It may, therefore, be that some of the genera only represent different growth forms in certain environments. In nature as well as in cultures, the sheaths of these

TABLE XIX. Examples of Solubilization of Metal Ores by Fe^{3+}.

Chalcopyrite:	$CuFeS_2(s) + 4Fe^{3+} + O_2 + 2H_2O \rightarrow Cu^{2+} + 5Fe^{2+} + 4H^+ + 2SO_4^{2-}$
Sphalerite:	$2ZnS(s) + 4Fe^{3+} + O_2 + 2H_2O \rightarrow 2Zn^{2+} + 4Fe^{2+} + 4H^+ + 2SO_4^{2-}$
Galena:	$2PbS(s) + 4Fe^{3+} + O_2 + 2H_2O \rightarrow 2Pb^{2+} + 4Fe^{2+} + 4H^+ + 2SO_4^{2-}$
Uranite:	$UO_2(s) + 2Fe^{3+} \rightarrow UO_2^{2+} + 2Fe^{2+}$

bacteria may become encrusted with deposited $Fe(OH)_3$ or MnO_2. In nature, the sheathed bacteria occur attached to substrates in running water such as springs, streams and water supply pipes with relatively low concentrations of dissolved organic materials, but they also occur in polluted rivers and streams and in sewage treatment plants. They do not occur in seawater. They often constitute an environmental nuisance, congesting water pipes and giving an unattractive appearance to streams. All strains studied grow on organic media (peptone, plant tissue infusions, etc.). In the presence of iron or manganese the bacteria precipitate $Fe(OH)_3$ or MnO_2 in their sheaths but the presence of these metals does not seem to stimulate growth. Attempts to demonstrate autotrophic growth Fe^{2+} or Mn^{2+} have generally been negative. Ali and Stokes (1971) reported increased cell yield in the presence of Mn^{2+} and growth in the absence of organic matter in strains of *Sphaerotilus*. These findings have been reinterpreted by van Veen (1972) as the result of protection against phosphate buffer, to which the organism is sensitive, in the presence of manganese, and also that the co-precipitation of protein with MnO_2 gave a false impression of cell growth in the absence of organic substrates (see also Dondero, 1975 for discussion). Rogers and Anderson (1976a, b) have given further evidence that the iron precipitation has no energetic significance for *Sphaerotilus* (colloidal Fe^{3+} is also precipitated) and that it is the non-living sheath material which somehow catalyzes the process. It has been suggested that the precipitation of iron or manganese serves as a detoxification mechanism against high concentrations of these metals (to which the bacteria are exposed in some of their typical environments, e.g., springs). However, evidence for this is not very convincing. Thus, while there is now strong evidence to show that the sheathed bacteria are not chemolithotrophs, the real physiological significance of the iron and manganese deposition remains unknown.

Among the iron-oxidizing bacteria the peculiar *Gallionella* genus should also be mentioned. *Gallionella* is a short, curved rod which produces a twisted stalk consisting of filaments of ferric hydroxides. Some evidence of autotrophic nutrition in *Gallionella* has been given; however, this can hardly be considered conclusive since pure cultures of the organism have not been studied (see Mulder, 1964). Together with the sheathed bacteria and some of the bacteria discussed below, *Gallionella* may mediate the deposition of lake or bog ore (limonite), viz., the precipitation of ferric hydroxides in acid lakes, bogs and springs where reduced iron is exposed to oxygenated water. The surface films consisting of ferric hydroxides often formed in bogs may also be the result of bacterial catalysis.

A number of other bacteria are also known to deposit ferric iron. These include *Siderocapsa* which deposits iron in mucoid capsules. Other bacteria which excrete mucoid polysaccharides (e.g., some *Bacillus* spp.) also tend to precipitate ferric iron. Strange filamentous bacteria belonging to the genus *Metallogenium* have especially been studied by Russian workers (e.g., Zavarin, 1964, 1968; Dubinina, 1970). *Metallogenium symbioticum* lives in some kind of association with a fungus; it accumulates MnO_2 in soils, a property it shares with many other soil bacteria. Other forms of *Metallogenium* are associated with iron and manganese deposition in aquatic environments. *Hyphomicrobium* is also capable of manganese oxidation in aquatic environments, (Tyler, 1970). In none of the above mentioned cases is chemolithotrophic oxidation indicated. In aquatic sediments, many forms of bacteria have the capability of catalyzing the oxidation of Mn^{2+}. Thus Schweisfurth (1973) found 30 strains from MnO_2-rich freshwater sediments which oxidize Mn^{2+} and Krumbein (1971) isolated many Mn^{2+} oxidizing bacteria (and fungi) from North Sea sediments.

Manganese nodules are layered, ferromanganese concretions which are widely distributed on the ocean floor and also in some large lakes. Due to a content of Ni and Cu, these concretions have attracted much interest within the last decade. Based on their content of radium, the growth rate of these nodules have been determined to be of the order of 1 mm per 1000 years. Their mode of formation is unknown but various evidence suggests a microbial participation in the process. Thus the nodules contain some organic material and Mn^{2+}-oxidizing bacteria have been isolated from them (Ehrlich, 1966; Zajic, 1969).

In conclusion, there can be no doubt that the bacterial oxidation of manganese and iron in soils and in aquatic environments constitutes significant processes from a geochemical point of view including the formation of certain types of ores. On the other hand the physiological significance of this widespread ability of heterotrophic bacteria remains largely elusive.

Microbial activity may also lead to the reduction of Fe^{3+} and Mn^{4+}. These may, in general, be considered to be processes which are indirectly mediated through a reducing or acid environment created by bacterial activity. In such a chemical environment created by bacteria in, e.g., water-logged soils or aquatic sediments, first MnO_2 and then $Fe(OH)_3$ and $FePO_4$ will solubilize and yield Mn^{2+} and Fe^{2+} respectively. Such reductions of these metals have been demonstrated in natural anoxic soils and waters and in laboratory experiments involving anaerobic bacteria (e.g., Bromfield, 1954; Zavarzin, 1968). Spontaneous reduction of colloidal $Fe(OH)_3$ and Fe^{3+}-organic complexes to Fe^{2+} in the oxidized surface layers of a bog, following a thermal mixing with the underlying anoxic water, has also been observed (Koenings, 1976). In anoxic groundwater, dissolved Fe^{2+} and Mn^{2+} will equilibriate with precipitated siderite ($FeCO_3$) rhodochrosite ($MnCO_3$) and calcite ($CaCO_3$) according to the pH value (Stumm and Morgan, 1970). In sediments where sulfate reduction takes place, the Fe^{2+}

and Mn^{2+} will precipitate as sulfides, being responsible for the black color of reduced aquatic sediments (see Section 6.2.1).

A few authors have found evidence that some bacteria may use certain metals in an oxidized state as a terminal electron acceptor. Woolfolk and Whitely (1968) found that in cultures of *Micrococcus lactilyticus* many different metals may serve as a hydrogen sink according to:

$$2Fe(OH)_3 + H_2 \rightarrow 2Fe(OH)_2 + 2H_2O.$$

The organism possesses hydrogenases and the process may well be of physiological significance by removing the hydrogen in a hydrogen-yielding fermentation similar to fermentative nitrate reduction (Section 5.3.1; see also Section 1.1.1). The significance of this sort of process for the reduction of metals is not known. Trimble and Ehrlich (1968, 1970) found evidence of heterotrophic bacteria using MnO_2 as a terminal electron acceptor. The physiological and ecological significance of these observations, however, remains unclear. The fact that the process was only initiated in aerobic environments and could take place in the presence of oxygen makes the interpretation of these findings questionable.

The last subject to be discussed briefly in this section is the microbial corrosion of iron. Since elemental iron is very rare in nature, this process is not a part of natural element cycling but will be mentioned here due to its practical importance. Different mechanisms may be involved in the corrosion of iron including acid formation (sulfur oxidation, acid fermentations), and the creation of galvanic cells by attached masses of bacteria (e.g., *Sphaerotilus*) on iron structures. The bacterial colonies form local reducing environments below them, and sulfate reducers utilize hydrogen at iron surfaces, thus depolarizing them. Hydrogen sulfide is also itself a corrosive agent. The microbial corrosion of iron is discussed in detail by Zajic (1969).

7.1.3. Other Metals

We have already seen that chemical transformations of some metals are mediated indirectly by changes in the pH or Eh of the environment brought about by bacterial activity. A general treatment of this belongs to environmental chemistry. It involves considerations of chemical equilibria and kinetics rather than microbial ecology; the reader is especially referred to Stumm and Morgan (1970).

Bacteria may also transform metals more directly. All organisms take up and excrete a large number of different elements in different forms. Several metals (e.g., Zn, Cu, Mg, Mo, Co) are components of one or more enzymes. These metals are concentrated by bacteria and, in certain cases, scarcity of these metals may limit microbial growth. This has, for example, been observed in the case of Mo in soils; Mo is necessary for the synthesis of nitrogenase and poor development of *Rhizobium* root nodules due to insufficiency of this metal has been recorded. Many metals form complexes with organic compounds. Bacteria and other organisms, therefore, often accumulate metals which are useless or toxic and organic excretions may chelate metallic ions. Bacteria also have detoxification mechanisms which transform toxic compounds of certain elements into less toxic ones. Finally, the literature contains claims that some compounds of more exotic metals may serve as either hydrogen acceptors or hydrogen donors in energy yielding processes of bacteria. For example, certain bacteria may mediate the oxidation of arsenite to arsenate and other forms have been shown to mediate the opposite process. Similarly the oxidation of Se° to $H_2SeO_3^-$ and the reduction of $H_2SeO_3^-$ to Se° may take place in bacterial cultures and have been claimed to be a normal part of the energy metabolism of these organisms. The real significance of such processes is, however, often very difficult to demonstrate and bacterial mediation has in some cases been inferred for processes which will take place spontaneously (cf. the discussion on manganese and iron oxidation in Section 7.1.2). Under all circumstances, the above mentioned processes must be trivial from a quantitative point of view. For general reviews on bacterially mediated transformations of metallic compounds see Silverman and Ehrlich (1969), Zajic (1969), and Jernelöv and Martin (1975).†

One of the aspects mentioned above and which may be of practical significance is the bacterial detoxification of certain metallic compounds. The metals in question occur naturally in small amounts in, e.g., seawater, and bacteria have evolved mechanisms in order to detoxify them. In cases where such metals are added to natural waters in industrial waste or from mining, the understanding of bacterial transformations of such compounds may be of paramount importance for predicting the consequences. Mercury may serve as an example. This metal occurs in three valence states: Hg°, Hg_2^{2+}, and Hg^{2+}. In anaerobic sediments mercury ions precipitate as HgS; this compound may be solubilized by sulfide oxidizing bacteria. Some bacteria are capable of reducing Hg^{2+} to Hg° using NADH. This is interpreted as a detoxification process as the formed elemental mercury eventually becomes volatilized. Some bacteria detoxify inorganic mercury species by methylation. The formation of methylmercury is of special interest due to its high toxicity for humans and its ability to accumulate in fish and molluscs. Finally some bacteria can transform CH_3Hg^+ into CH_4 and Hg° (Fig. 41). The actual concentration of methylmercury in an aquatic system is, therefore, a function of a number of processes which again depend on various environmental parameters (pH, Eh, etc.) and it therefore does not directly reflect the rate of

† See also recent review by H. L. Erlich (1979), *Geomicrobiol. J.* **1**, 65–83.

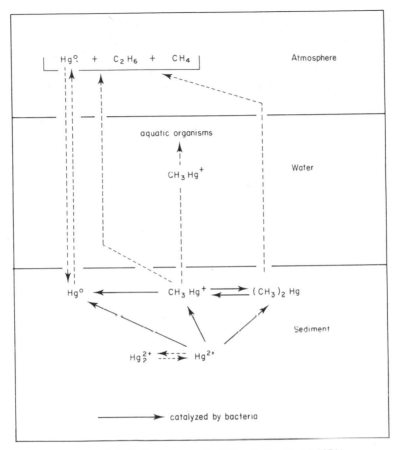

Fig. 41. The biological mercury cycle. (Altered after Wood, 1974.)

its production. Other metals, e.g., arsenic, undergo comparable biological cycles in nature (see Wood, 1974 and references therein).

7.2. PHOSPHORUS

In nature phosphorus occurs nearly exclusively as phosphates in which the P-atom has an oxidation state of $+5$ and it does not change valence states in natural, biological, or abiological chemical transformations. (In reducing soils, clostridia and other anaerobes are reported to mediate the reduction of HPO_4^{2-} to HPO_3^{2-} and HPO_2^{2-}; for references see Silverman and Ehrlich, 1964.)

In natural waters and in soils some of the dissolved phosphorus is in the form of orthophosphate (HPO_4^{2-} and $H_2PO_4^-$ according to the hydrogen ion concentration). It derives from the natural weathering and erosion of phosphate minerals, the solubilization of precipitated metallic phosphates and adsorbed phosphate, and the excretion of bacteria and other organisms; to these sources can be added soil fertilizers and industrial and domestic waste water. Another fraction of dissolved (and in part colloidal) P consists of organic phosphate esters. This fraction, which typically constitutes between 15 and 60% of the dissolved phosphorus, is chemically poorly defined; it is derived from living organisms in the form of excreta or leachates and from the autolysis of dead organisms. Polyphosphates (i.e., condensed ortho-phosphates) of varying molecular weight which are synthesized by living organisms also constitute a part of the pool of P in natural water. Typically the total pool of dissolved P of natural waters varies between 5 and 1000 μg $P l^{-1}$; of this pool, however, only a part, notably the orthophosphate fraction, is readily taken up by organisms.

Phosphates have strong tendency to be adsorbed to clays and to form insoluble metallic salts, in particular with Ca^{2+}, Fe^{3+} and Al^{3+}. Since P does not appear in a gas phase in nature, its return to terrestrial environments from the sea is quantitatively insignificant (sewater spray and guano). As P is continuously transported from land to the sea with rivers, there is a net precipitation of phosphates in the sea which acts as a global phosphorus sink until the phosphates, in the form of various minerals (e.g. apatite), are returned through mountain building over geological time. For a general treatment of the chemistry of phosphorus in natural waters and soils, see Stumm and Morgan (1970) and Cosgrove (1977).

Bacteria (and algae and higher plants) take up dissolved orthophosphate from the water. Organic phsophates may be taken up directly or they are first hydrolyzed by extracellular alkaline phosphatases, but organic P may be very resistant to hydrolysis and not readily available to microorganisms. In the cell, orthophosphate is coupled to ADP to form ATP. Phosphate is essential for the transfer of energy and phosphorylations, and for the synthesis of nucleic acids in all living cells. Bacteria (and some other organisms) may store phosphorus in *volutin granules*; this pool may make up a considerable fraction of the total cell phosphorus. The surplus phosphorus is in the form of linear polyphosphates of various lengths. It is especially formed when the bacteria are limited by some other nutrient (e.g., sulfate). When the cells are again allowed to grow, the phosphates are rapidly mobilized for DNA synthesis (Rose, 1968; Zajic, 1969; Stanier *et al.*, 1970; Barsdate *et al.*, 1974; see also Fig. 42).

In natural waters and soils the amount of P contained in living organisms usually far exceeds the amount of dissolved orthophosphate in the environment and the exchange rate between these two pools is much greater than the

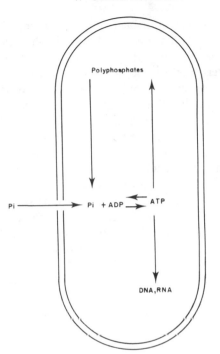

Fig. 42. A simplified representation of the pools of phosphorus in a bacterial cell.

net deposition of undissolved phosphorus or leaching from soils. The turn-over time of orthophosphate may be very short (viz., down to 2 min) in connection with a high biological activity (algal blooms, bacterial growth on easily decomposed organic material, etc.). Where the biological activity is low or where there is an unusually high orthophosphate concentration, the turnover time may exceed 100 hours.

The biological phosphorus cycle is very simple in principle. The essential element is taken up and used in the cell as phosphate and it is also liberated by excretion or through the mineralization of dead organic matter as phosphate. In this respect bacteria do not play any qualitatively different role from that of other organisms. Just like algae and higher plants, they have evolved efficient uptake mechanisms and the growth of biomass may be limited by the availability of phosphate in the case of all these organisms (Medveckzy and Rosenberg, 1971; Fuhs *et al.*, 1972; Rhee, 1972; Barsdate *et al.*, 1974; see also Section 3.3).

The interesting biogeochemical aspect of P is principally that due to its tendency to form insoluble compounds and organic phosphate esters which are very resistant to hydrolysis, it is often the factor which limits ecosystems. The special role of bacteria in the biological phosphorus cycle is (in addition, of course, to their role in mineralization) that they may indirectly mediate the release of dissolved orthophosphate from insoluble phosphates. Calcium phosphates are dissolved at low pH values and acid producing bacteria are believed to play a significant role in the release of orthophosphate from phosphate minerals in soils. It has been the practice to add elemental sulfur to fertilizers thereby stimulating acid production by thiobacilli. In tropical soils this does not seem to improve phosphate availability (Cosgrove, 1977).

In aquatic systems the sediments act as a phosphorus trap. This is in part due to the sedimentation of particulate, dead organic matter which is mineralized in the sediment and in part due to the precipitation of metallic phosphates, especially $FePO_4$. In anaerobic sulfide-containing sediments, this salt is reduced, releasing the orthophosphate. There is, therefore, a net flow of dissolved phosphate from the sediment to the overlying water, a flow which, however, has proven difficult to quantify. If an oxidized sediment surface layer is developed, the release of orthophosphate is limited since it tends to reprecipitate with ferric iron as soon as it reaches the aerobic zone. On the other hand, if the reducing zone reaches the sediment surface, larger

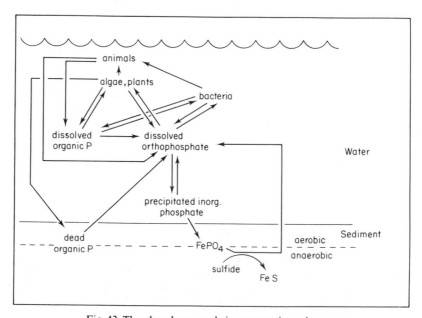

Fig. 43. The phosphorus cycle in an aquatic environment.

amounts of orthophosphate are released to the water. This mechanism is very important in stratified lakes. In periods with an anoxic hypolimnion, large amounts of orthophosphate accumulate in the hypolimnion which after the spring mixing becomes available to the plankton and algae in the epilimnion. The mechanism is also important in the process of lake eutrophication, e.g., as the result of addition of water from sewage treatment plants. As long as the surface sediments are oxidized, the sediments can accumulate large amounts of phosphorus with restricted adverse effects. When the productivity of the lake reaches a point where the sediment surface or even the bottom waters become anoxic and sulfide-containing, this phosphate is again released, however, enforcing the eutrophication process (Mortimer, 1941–42; Stumm and Morgan, 1970). A general presentation of the P-cycle in an aquatic system is shown in Fig. 43.

CHAPTER 8

Microbial Symbiosis and Mineral Cycling

The adaptive significance of symbiosis (here meant to describe any spatially or physiologically intimate "living together" and thus including mutualism, parasitism and commensalism) is varied and may comprise transport, protection or exotic phenomena such as luminescent bacteria in certain species of fish and squids. However, most cases of symbiosis involve an one-sided or reciprocal transfer of nutrients and the basis of the association is often the possession of complementary metabolic or biosynthetic pathways. All natural eco-systems provide many examples of symbiosis, several of which await a full understanding of their significance. In this chapter we will only discuss a few cases of symbiosis which involve prokaryotes and which are important in the transformation of matter in nature or in other ways illustrate microbial element cycling. No attempt to give a general review of symbiotic prokaryote–eukaryote relationships will be made.

Mutualistic symbiosis is not always very sharply defined. In all eco-systems some organisms utilize the metabolic products of other species, e.g., photoautotrophic and heterotrophic organisms, sulfate reducers and sulfide oxidizers; other relationships, such as the synthesis of vitamins for other species, may be more subtle. The degree of intimacy of such interdependences varies and this poses terminological difficulties. Endosymbionts are not found in bacteria. However, several close relationships between two kinds of bacteria have been described although the nature of the associations is often not understood. Many blue-green bacteria are covered by hetero-trophic bacteria, the latter presumably utilizing the evolved oxygen and/or excreted organic material. The co-occurrence of cellulolytic bacteria (*Cyto-phaga*) with non-cellulolytic forms (e.g., *Sphaerotilus*) is frequently seen but the nature of the association is not fully understood. This also applies to *Chlorochromatium* which is a consortium consisting of a large, flagellated bacterium covered by usually 6 or 8 cells of a green sulfurbacterium. An interesting example of mutualism is constituted by "*Methanobacillus omeli-anskii*", an organism which in cultures oxidizes ethanol to acetate with CO_2 as a hydrogen acceptor according to:

$$2CH_3CH_2OH + CO_2 \rightarrow 2CH_3COOH + CH_4.$$

It has been shown, however, that "*Methanobacillus omelianskii*" is in fact a mixed culture consisting of two species. One of these catalyzes the process:

$$CH_3CH_2OH + H_2O \rightarrow CH_3COOH + 2H_2$$

and the other species is a methanogenic bacterium catalyzing the process:

$$4H_2 + CO_2 \rightarrow CH_4 + 2H_2O,$$

(Bryant *et al.*, 1967). The free energy yield of the former process depends on the removal of the hydrogen formed, thus making the two bacteria mutually interdependent (see also Section 1.1.1). Hydrogen transfer between anaerobic bacteria is important in the rumen system (discussed in the following section) and presumably also in anaerobic soils and sediments (Section 4.1.5). A similar "syntrophism" between a sulfur-reducing *Spirillum* and *Chlorobium* was recently described by Wolfe and Pfennig (1977).

The eukaryotic cell may have originated from symbiotic relationships between prokaryotes which then has had immense evolutionary significance. This aspect will be discussed in Section 8.4. We will, however, first discuss two well studied types of symbiotic relationships involving bacteria and which are of significance for element cycling, viz., the symbiotic degradation of structural carbohydrates and symbiotic nitrogen fixation.

8.1. THE SYMBIOTIC DEGRADATION OF STRUCTURAL CARBOHYDRATES

As already mentioned in Chapter 3, animals in general are not capable of digesting structural plant compounds. However, different taxonomic groups have independently evolved symbiotic relationships with microorganisms which allow them to utilize, e.g., cellulose.

The best studied case, i.e., the microflora of ruminant animals, has also yielded a general insight into fermentation processes in a natural system; in fact our discussion on fermentation in anaerobic sediments is in part based on information from the rumen system. We will, therefore, discuss it in some detail.

8.1.1. The Rumen System

The detailed knowledge of this system is mainly due to the work of Hungate and his collaborators (e.g., Hungate, 1963, 1966, 1975; see also Dougherty, 1965). In a ruminant mammal the ingested feed, possibly regurgitated and rechewed, passes from the *reticulum* (Fig. 44) to the *rumen*, together with saliva which is secreted more or less continuously. The rumen which comprises up to 15% of the volume of the animal may be considered as a continuous fermenter. The content of the rumen is strictly anoxic and reducing (Eh ~ −350 mV); the bicarbonate of the saliva buffers the pH value to around 6·5. In the rumen, the carbohydrates of the feed are microbially fermented into methane, carbon dioxide and fatty acids. The contents then pass from the rumen to the *omasum* where the fatty acids and bicarbonate are absorbed, the fatty acids being the main energy source of ruminants (some absorption also takes place in the rumen). In the following section of the digestive system, the *abomasum* (the true stomach), acid digestion of the microbial cells takes place; the microbial cells constitute the protein component of the animal's food. The turnover of the rumen content is of the order 1–1·5 per day; typically a ruminant spends one third of its time eating. The rumen contains up to 10^{11} bacteria and 10^5–10^6 protozoan cells per ml; the animal is totally dependent on this microbial community for conversion of the ingested food into compounds it can utilize.

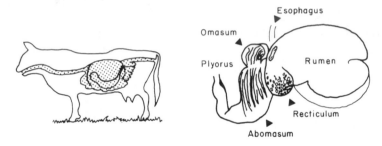

Fig. 44. The digestive system of a ruminant.

The overall stoichiometric equation for the fermentation of hexose in the rumen can be given as:

$$58C_6H_{12}O_6 \rightarrow \underset{\text{acetic}}{62CH_3COOH} + \underset{\text{propionic}}{22CH_3CH_2COOH} +$$

$$+ \underset{\text{butyric}}{16CH_3CH_2CH_2COOH} + 60\cdot5CO_2 + 33\cdot5CH_4 + 27H_2O.$$

This equation gives a realistic presentation of the ratio of fermentation products actually found in most cases. The end products are the result of several interacting microbial processes. They involve a number of intermediate products which only occur in minute amounts but have a rapid turnover. We will first discuss the primary degradation of the compounds ingested with the food and then proceed to discuss the fermentation processes and the intermediate products. (We will here restrict ourselves to a discussion of the bacterial components of the microbial communities. The protozoa, i.e., flagellates and in particular, ciliates, constitute an important aspect of rumen ecology; however, it is possible to discuss the main pathways of carbon in the rumen without including them.)

The feed of ruminants consists of plant material and thus carbohydrates constitute the bulk of the ingested material. Celluloytic bacteria account for only 5–15% of the bacterial cells in the rumen. Considering the typically high cellulose content of the food, this seems a surprisingly small figure. The explanation given for this is that the cellulolytic bacteria hydrolyze more cellulose than they utilize, the excess sugar being utilized by other sugar-fermenting bacteria. Among the cellulolytic bacteria can be mentioned *Bacteroides succinogenes*, a gliding form related to *Cytophaga*. It hydrolyzes cellulose and ferments glucose and cellobiose (and some strains also other sugars) into mainly succinic but also acetic and formic acids. Other bacteria which degrade cellulose (and other polysaccharides) are *Butyrivibrio fibrisolvens, Clostridium lochheadii, Ruminicocci* and some other rods and cocci. The various cellulolytic bacteria differ with respect to the presence of exoenzymes. Many of the cellulolytic bacteria are probably attached to plant particles in the rumen. The cellulolytic activity is believed to depend on synergistic relations with other bacteria. This may in part be due to the breaking of various other chemical bonds exposing cellulose and also since some of the cellulose decomposers require volatile acids synthesized by other forms. Starch is degraded by several rumen bacteria; in addition to some of the above mentioned cellulolytic forms, *Bacteroides amylophilus, B. ruminicola, Streptococcus bovis, Succinomonas amylotica* and *Selenomonas ruminatium* may be mentioned. Hemicelluloses (xylan, pectin, gum arabic) are degraded by many rumen bacteria including several of the above mentioned forms and species of *Ruminococcus*. Altogether, carbohydrates are very efficiently degraded in the rumen. This is in contrast to lignin which is not attacked at all; in fact, it has been used as a conservative component in order to estimate the assimilation efficiency of ruminants.

Proteolytic bacteria can also be demonstrated in the rumen. Proteins included in the food are rapidly fermented, leading to the production of NH_3. In addition to the organic nitrogen of the food, the rumen is supplied with urea excreted with the saliva. In the rumen the urea is rapidly fermented into NH_3 and CO_2. When the ruminant eats protein-deficient feed, the nitrogen supplied with the urea is an important supplement allowing bacterial protein synthesis to occur. The rumen microorganisms seem preferentially to use NH_3 as a nitrogen source and amino acids occur in extremely low concentrations. Due to the rapid fermentation of proteins in the rumen, the addition of proteins or amino acid to the feed of cattle is not reflected in a corresponding increase in the nitrogen retention of the ruminant which is almost totally dependent on the protein synthesized by bacteria.

TABLE XX. Fermentation Products of Rumen Bacteria.[a]

	Formate	Acetate	Propionate	Butyrate	Higher acids	Lactate	Succinate	Ethanol	CO$_2$	H$_2$	CH$_4$
Bacteroides succinogenes	x	x	—	—	—	—	x	—	u	—	—
Ruminococcus flavefaciens	x/-	x	—	—	—	x/-	x	x/-	u	x/-	—
Ruminococcus albus	x	x	—	—	—	x/-	x/-	x	x	x	—
Bacteroides amylophilus	x	x	x	—	—	x/-	x	x	u	—	—
Succinomonas amylotica	—	x	x	—	—	—	x	—	u	—	—
Veillonella alcalescens	—	x	x	—	—	—	—	—	x	x	—
Methanobacterium ruminantium	—	—	—	—	—	—	—	—	u	u	x
Anaerovibrio lipolytica	—	x	x	t	x	—	x	—	—	—	—
Peptostreptococcus elsdenii	—	x/u	x	x	—	—	—	x	x	x	—
Clostridium lochheadii	x	x	—	x	—	x/-	—	x	x	x	—
Clostridium longisporum	x	x	—	—	—	x/-	—	x	x	x/-	—
Borrelia sp.	x	x	—	—	—	—	x	—	u	—	—
Lachnospira multiparus	x	x	—	—	—	—	x	x	x	x	—
Ciliobacterium cellulosolvens	x/-	x/-	x	x/-	—	x	x/-	—	—	x	—
Butyrivibrio fibrisolvens	x	x	—	x	—	x	—	x	x	x	—
Butyrivibrio alactacidigenes	x	x	—	x	—	—	—	x	—	x	—
Bacteroides ruminicola	x	x	—	—	—	—	x	—	u	—	—
Selenomonas ruminantium	x/-	x	x	x/-	—	x/-	x/-	x	—	—	—
Selenomonas lactilytica	—	x	x	—	—	x/-	x/-	—	—	—	—
Succinivibrio dextrinosolvens	x	x	—	—	—	x/-	x	x	u	—	—
Streptococcus bovis	x/-	x	—	—	—	x	—	x/-	x/-	—	—
Eubacterium ruminantium	x	x	—	x	—	x	—	—	x	—	—

[a] After Hungate (1966). x/-: produced by some strains; u: used; t: trace.

Table XX shows the fermentation products of various rumen bacteria in pure culture. It can be seen that these products are more diverse than the quantitatively important end products of the rumen fermentation: butyric, acetic and propionic acids and methane and carbon dioxide. The reason for this is that the other products enter into other fermentation processes in the rumen (Table XXI).

The efficient removal of H_2 through methanogenesis catalyzed by *Methanobacterium ruminatium* is a key process in the rumen. In lactic and ethanol fermentation, the pyruvate formed through glycolysis is used to reoxidize NADH thus producing lactate or ethanol plus CO_2. When the hydrogen pressure is kept sufficiently low, however, the process $2NADH \rightarrow 2NAD^+ + H_2$ is exergonic and lactate and ethanol production is, therefore, low in the rumen. The small amounts of lactate and ethanol actually produced are further fermented to propionic and acetic acid by, e.g., *Micrococcus lactily-ticus*. Most of the pyruvate formed by glycolysis is, therefore, either utilized for propionic acid fermentation or it is split into acetyl-Co A, CO_2 and H_2. The latter process is responsible for most of the hydrogen generation in the rumen. Acetyl-Co A is then fermented into butyric or acetic acid. These processes will yield some ATP in addition to the ATP generated by glycolysis. Thus, the removal hydrogen by methanogenesis allows for a higher cell yield than would lactic fermentation, in addition to the yield of methanogenic bacteria.

Some of the pyruvate is split into acetyl-Co A and formate rather than acetyl-Co A, CO_2 and H_2. Formate then decomposes into CO_2 and H_2: this process is believed to be responsible for about 18% of the H_2 generated in the rumen. It is also possible that some of the formate is oxidized with malate or fumarate thus forming succinate, a process which may be catalyzed by *Vibrio succinogenes* (Hungate, 1966).

Many of the bacteria with a propionate fermentation pathway seem to release succinate rather than propionate in pure cultures (e.g., *Bacteroides*

TABLE XXI. Turnover of Intermediates in Rumen Fermentation.[a]

	Concentration $(nmol\,ml^{-1})$	Turnover rate (min^{-1})	Flux $(nmol\,ml^{-1}\,min^{-1})$	Product formed	Rumen production accounted for (%)
Lactate	12	0·03	0·36	propionate	?
				acetate	1
Ethanol	trace	0·003		acetate	?
Succinate	4	10	40	propionate	33
H_2	1	710	710	methane	100
Formate	12	10	126	H_2	18

[a] After Hungate (1975).

succinogenes and *B. amylophilus*). When succinate is added to rumen contents it is converted into propionic acid and CO_2, a process which can be performed by *Veillonella gazogeneous*. It has been speculated that the failure of many bacteria in the rumen to decarboxylate succinate into propionate may be due to the high availability of CO_2 in the rumen environment. Blackburn and Hungate (1963) showed from pool sizes of succinate, turnover rates and the total production of propionate, that the decarboxylation of liberated succinate accounts for about 33 % of the production of propionate in the rumen (Table XXI). The main fermentative pathways (based on sugars) are shown in Fig. 45.

If a cow, previously fed with a poor food, is suddenly allowed to feed on starch-rich food, lactic acid fermentation, carried out by *Streptococcus bovis*, may dominate. This leads to an accumulation of lactic acid and a decreasing pH, both of which inhibit the normal microbial community and may even result in the death of the animal. If the cow is adapted more slowly to a starch-rich diet, this will not happen. The underlying reason for these obser-

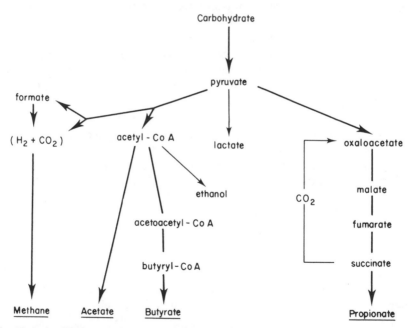

Fig. 45. A simplified representation of the main fermentative pathways of carbohydrate in the rumen. The methanogenic pathway is catalyzed by *Methanobacterium*. The other main pathways leading to acetate, butyrate and succinate and propionate cannot unambiguously be assigned to individual bacterial species but rather indicates the main pathways of the carbon flow in the rumen. The relatively small amount of lactate formed is fermented into propionate and acetate, and ethanol is believed to be transformed into acetate.

vations is that the microorganisms in the rumen are normally energy limited. The fermentative pathways giving the highest yield of ATP per substrate unit, viz., the pathways leading to butyrate, acetate or propionate are competitively superior. The small amounts of lactate produced under normal conditions are readily channelled through one of the above mentioned fermentative pathways by other bacteria. However, if substrate is suddenly in excess, the lactic acid bacteria, with a lower yield of ATP per mol hexose, but with a higher specific growth rate are competitively superior. The normal bacterial flora cannot cope with the lactate produced rapidly enough to restore normal conditions. Normal lactate formation and utilization has been well studied in ruminants since silage has a high lactate content. Most lactate seems to be fermented into propionate, the proportion of which increases relative to acetate in silage fed cows (Walker, 1968).

All the volatile acid end products serve as an energy source for the ruminant. However, while acetate and butyrate are either metabolized or stored as fats, propionate is the only end-product from which the animal can readily synthesize carbohydrates. The inhibition of methanogenesis (which can be done experimentally by administering chloroform to the rumen) should lead to an increased partial pressure of hydrogen which again should increase the yield of propionate relative to butyrate and acetate. Experiments have shown this prediction to hold true; there are even reports which indicate that the treatment is beneficial to the growth of cattle (Hungate, 1975).

The anaerobic nature of the rumen system has important energetic implications. As already mentioned, the microbial community is normally energy limited. In the rumen the yield of ATP per mole hexose is about 4 (as compared with an aerobic oxidation which would yield 38 ATP). This corresponds to a yield of about 40 g bacteria per mole hexose (Section 1.1). The remaining energy of the digestible material is conserved in CH_4 (about 10%, it is belched by the ruminant and represents a loss) and the volatile fatty acids ($\sim 65\%$), the energy of which is available for the animal since it is an aerobe. While the microbial community, as already mentioned, is energy limited, the ruminant is mainly limited by the availability of nutrients, in particular nitrogen, which is met by microbial protein. The growth efficiency of ruminants is, therefore, relatively low. The competitive advantage of ruminants compared to other mammals in nature (and as domestic animals) is that, through the strange and complex symbiotic relationship with microorganisms, the exploitation of a resource inaccessible to most other animals is possible.

Although the rumen system may be considered the best understood example of a community of anaerobic microorganisms or perhaps even the best understood ecosystem, a number of questions are not yet answered. Thus, it is not completely understood which fermentative pathways are selected, although it is obviously a very important problem, not least from the point of view of the ruminant, as the situation of excessive lactate fermentation mentioned above shows. One may, for example, ask why acetate is not fermented into CO_2 and CH_4, a process which is carried out by some bacteria (Section 1.1.1). This would obviously lead to an energy loss for the ruminant. An important factor in the quantitative regulation of the different pathways is probably the turnover time of the rumen contents in conjunction with different specific growth rates of the bacteria involved (see Section 4.1.3 and Fig. 25). A complete description of the rumen system including such considerations has not yet, however, been achieved.

8.1.2. Non-Ruminant Associations

The ruminants constitute the best studied example of a symbiotic association between carbohydrate-fermentating microorganisms and herbivores but it is by no means the only one. Many other mammal groups have independently developed similar systems in a more or less specialized stomach or in an enlarged cecum. Examples are provided by certain marsupials, edentates and rodents, and also by lagomorphs, elephants and perissodactyls. Recently such a system has also been demonstrated in the marine turtle *Chelonia mydas* which feeds on seagrasses (Fenchel *et al.*†). All these forms harbor bacteria which hydrolyze ingested structural carbohydrates and ferment them to volatile fatty acids. Species with a cecal fermentation (e.g., perissodactyls) are, in one sense, at an advantage since they can utilize the low molecular weight carbohydrates and proteins of the feed directly, prior to the bacterial fermentation (Janis, 1976). The ruminants, however, are usually considered to have evolved the most specialized and efficient system. This interpretation is supported by the evolutionary history of the ungulates ("hoofed animals"). The perissodactyls appear in the beginning of the Eocene; during this and the following Oligocene period they constitute (together with some extinct groups) the dominating group of large herbivore mammals. The artiodactyls also appeared in the beginning of the Eocene. Within this group, the true ruminants (excluding camels) originated in the Oligocene. During the following Miocene period and up to the present time, the ruminants have shown a rapid adaptive radiation and have, over a relatively short time span, displaced other mammals to become the dominating large herbivores. Of 158 known genera of perissodactyls there are only six living ones, whereas there are 73 living genera of ruminants (Romer, 1958; Moir, 1965).

Within arthropods a great number of associations with microorganisms (fungi, protozoa, bacteria) are known. The presence of such associations

† *Appl. Environ. Microbiol.* (1979) **37**, 348–350.

correlate well with the host diet, viz., they mainly occur in forms feeding on cellulose-rich plant material, keratin and similar structural, nutrient-poor substrates (Brooks, 1963). Most of these relationships have, unfortunately, not yet been studied in detail. Some of the best studied and most interesting examples include the relationship between wood-boring beetles and fungi. In wood-eating cockroaches and termites the degradation of the ingested food is also due to symbiotic microorganisms. In one of the five families of termites, the microorganisms are exclusively bacteria. In other families and in the cockroaches, a strange fauna of flagellates is the most obvious feature of the digestive system. However, even in this case, bacteria are present, as endo- and ecto-symbionts of the flagellates and also free-living in the gut. It is therefore possible that these bacteria are also responsible for the cellulose degradation. The specialized attachment sites in the termite gut for bacteria suggest that they play an important role (Breznak and Pankrantz, 1977). The termite gut is anoxic like the rumen and carbohydrates are fermented into acetic acid and other volatile fatty acids which constitute the energy source of the termite (Hungate, 1955). The phylogenetic history of some insect groups, in conjunction with the present day distribution of symbiont groups, indicates that these relationships had already evolved in the Carboniferous period (Brooks, 1963).

Sea urchins feeding on macroalgae have very dense populations of alginolytic bacteria in their intestinal tract (Lasker and Giese, 1954; Prim and Lawrence, 1975). While it has not been shown directly, this flora probably plays a role for the host comparable to that in herbivorous mammals and insects.

There are a few examples of animals producing cellulases themselves (viz., the wood boring isopod *Limnoria* and the pulmonate snail, *Helix*). Still it is safe to say that by far the greatest part of animal herbivory on higher plants is accomplished by a rather restricted number of animal taxons, which have developed symbiotic relations with microorganisms. In most ecosystems, the fraction of structural plant compounds mineralized by such herbivore animals is still quite small (see Section 3.1). From the viewpoint of zoology and general ecology, however, the evolution of the symbiotic relationship between herbivores and microorganisms is of great significance.

8.2. SYMBIOTIC NITROGEN FIXATION

The reduction and assimilation of dinitrogen is a unique property of certain prokaryotes. While symbiotic cellulose degradation is probably of a restricted importance for the global carbon cycle, the symbiotic nitrogen fixation is quantitatively important for the global nitrogen cycle. Biological

N_2-fixation is the most important pathway leading from atmospheric N_2 to biologically bound N (see Section 9.2). Estimates are still very inaccurate but it is currently believed that between 15 and 22 % of all biological nitrogen fixation is carried out in symbiotic associations among which the legume–rhizobia association is probably the single most important type (Quispel, 1974). There are two important prerequisites for N_2-fixation, viz., a high availability of energy and a low oxygen tension. These requirements are provided by the host organisms, explaining the importance of symbiotic nitrogen fixers relative to free-living ones.

The study of nitrogen fixation has recently attracted much interest and, therefore, has yielded much new information. There are two reasons for this. The development of new techniques (viz., the use of ^{15}N labeling and the acetylene-reduction method for measuring nitrogen reduction) has made the quantification of nitrogen fixation of whole soil samples, of entire organisms or of *in vitro* systems relatively easy. Also, the high price in terms of energy of industrially produced fertilizers has increased the incentive to optimize the use of biological nitrogen fixation. All these efforts are reflected in several recent reviews and memoirs on the subject (e.g., Quispel, 1974; Burns and Hardy, 1975; Nutman, 1976; Brill, 1977) on which this section is mainly based. The most important and well studied cases of symbiotic nitrogen fixation are compiled in Table XXII.

TABLE XXII. Well Established Examples of Symbiotic N_2-fixation.

Associated organisms	N_2-fixing symbionts	Symbiont-containing organs
Legumes	*Rhizobium* spp.	root nodules
Alnus, Myrica, Dryas, Casuarina, Elaeagnus, Hippophaë, etc. (woody angiosperms)	"Actinomycete like" bacteria	root nodules
Codium fragile (marine alga)	*Azotobacter?*	on surface of thallus
Azolla (fern)	*Anabaena azollae*	cavity in leaves
Cycads	*Nostoc, Anabaena*	coralloid roots
Gunnera (angiosperm)	*Nostoc punctiforme*	glands in stems and rhizomes
Collema, Leptogium, and other lichens	*Nostoc, Calothrix,* and other blue-greens	surrounded by hyphae
Termites	*Citrobacter freundii Enterobacter agglomerans*	gut

8.2.1. N$_2$-Fixation in Legumes

The root nodules of legumes constitute the best studied and most highly evolved case of symbiotic nitrogen fixation. It may also be the quantitatively most important system; this is certainly so when considering agricultural practice. The dinitrogen reduction is carried out by soil bacteria belonging to the genus *Rhizobium*. These Gram-negative rods are heterotrophs which can utilize a variety of organic substrates as energy and carbon sources. In cultures (except under special conditions) and in the soil they cannot utilize dinitrogen but require some combined (organic or inorganic) nitrogen source. They are usually found in low numbers in soil but their growth is stimulated in the rhizosphere of legumes.

The relation with legumes is a very specific one, that is, each legume has its own *Rhizobium* species. It has recently been shown for soybeans and for clover that this may be related to a protein excreted by the plant. These proteins (named *trifoliin* in the case of clover) have identical binding sites on the root hairs (where the primary infection takes place) and on the surface of the specific *Rhizobium* sp. This leads to a binding of the bacteria to the root hairs of the right host plant (Brill, 1977).

The specificity and low numbers of rhizobia in soils have led to the practice of seed inoculation with rhizobia where a given legume species has not previously been grown in a soil.

Infection takes place from the tips of the root hairs. The rhizobia induce the hair to make an *infection thread* starting as an invagination of the cell wall. The infection thread grows backwards and eventually breaks to liberate the rhizobia into a specialized cortical cell. This initiates the nodule formation which involves a number of cellular specializations controlled by phytohormones. It was earlier believed that the intracellular bacteria are found in the cytoplasm of the host cells; electron microscopic observations indicate, however, that they are enveloped by cell membranes and may be considered as being phagocytized. Initially the bacteria divide within the host cell but as the nodule develops they transform into non-dividing *bacteroids*, i.e., they swell and become irregular in shape. These bacteroids are now capable of nitrogen fixation. It has recently been shown that isolated and purified bacteroids, and also the dividing stages of rhizobia prior to the bacteroid formation, are capable of nitrogen fixation outside their host, when given a low oxygen tension.

After some time, the nodules degenerate and the bacteroids lyse. The degenerating nodules, however, are invaded by vegetative rhizobia which have remained in the infection thread; in the degenerating nodule they can grow and multiply.

One of the most striking host adaptations to this symbiosis is the production of a hemoglobin, *leghemoglobin*, in the nodules. It is the only hemoglobin known from the plant kingdom and it has the general property of reversible oxygenation. The function of leghemoglobin is to keep the oxygen tension low in the nodules, thus protecting the nitrogenase, and at the same time providing the bacteria with a hydrogen acceptor so that the bacteroids can maintain a high energy metabolism necessary for the dinitrogen reduction. For a detailed description of the physiology and development of legume root nodules see especially the reviews by Nutman (1963), Dart (1974) and Vincent (1974).

The legumes constitute a very species-rich family of plants and they have a world wide distribution. In agriculture the quantitative importance of N_2-fixation by the legumes is well documented. Between 20 and 500 kg N_2 ha^{-1} $year^{-1}$ have been reported, values of around 100 kg being most common. Estimates of the importance of nitrogen fixation by legumes in natural ecosystems and on a global basis are still very uncertain. Figures for the latter offered in the literature vary between 14 and 100×10^6 tons per year.

8.2.2. Other N_2-Fixing Plant—Eubacteria Associations

While the *Rhizobium* symbiosis is confined to the numerous genera and species within one family, another type of nitrogen fixing root nodule is distributed among 13 largely unrelated genera of woody plants, of which the best known example is the alder (*Alnus*). Our understanding of these associations is restricted by the fact that the symbiotic bacteria have so far not been grown in cultures outside their hosts.

The infection process and the nodule formation is well established. Infection takes place from the soil. Seeds planted away from parent stands often do not develop nodules. Cross infections are possible between species within the genus *Alnus* and between species within the family Elaeagnaceae. The general morphology and fine structural details of the symbionts suggest an affinity to the actinomycetes. The presence of hemoglobins has not been established; some cases of red colored nodules are due to the presence of anthocyanins.

The nitrogen fixation and its significance for the host plant has been demonstrated in many cases. Fixation rates of 100 to 200 kg N_2 ha^{-1} yr^{-1} have been found for growths of *Hippophaë* and *Alnus*. Many plants in this group are found in acid, waterlogged soils (e.g. *Alnus* and *Myrica*). Where they occur close to lakes or rivers they may contribute considerable amounts of fixed nitrogen to these freshwater systems. For a more detailed treatment of these non-legume associations, see Bond (1963, 1974) and Nutman (1976).

In addition to the complex and highly evolved examples of symbiotic nitrogen fixation discussed so far, less specific and complex associations have also been described. It has been demonstrated that the nitrogen fixation of free-living bacteria is stimulated in the rhizosphere where their population densities are also higher. This is probably an effect of root exudates. Nitrogenase activity can also often be demonstrated on the root surfaces of plants due to attached bacteria. It has also been found that some grasses have nitrogen fixing bacteria growing on the root surfaces. Root nodules do not form but apparently the host gains from the associations (Döbereiner, 1974; Day *et al.*, 1975; Brill, 1977). Recently Head and Carpenter (1975) have found that the marine macroalga *Codium fragile* is associated with *Azotobacter*-like bacteria. The N_2-fixation of these bacteria living on the surface of the thallus is coupled with the release of dissolved organics from the host.

8.2.3. Associations with Blue-Green Bacteria

Cyanobacteria are known to occur in symbiosis with many organisms; these associations are all potentially nitrogen fixing but this has not been studied in all cases. A number of protists (unicellular algae, flagellates, phycomycetes) have "endocyanosises", viz., intracellular blue-green bacteria. The nitrogen fixing ability in these forms has yet to be established. Symbiotic relationships between *Nostoc* and certain liverworts seem to be nitrogen fixing. Among the lichens only a minority of the species (about 8%) have cyanobacteria as the photosynthetic component. Several of these lichens (e.g., *Collema*, *Peltigera*) have been shown to fix nitrogen.

The association between the small aquatic fern, *Azolla*, and the blue-green, *Anabaena azollae*, has been studied in some detail. *Azolla* is found floating in fresh waters in tropical and subtropical areas all over the world. In South East Asia it is grown in flooded rice paddies and used as a green manure. Dinitrogen fixation by dense populations of *Azolla* has been estimated to values of between 100 and $600 \, kg \, N_2 \, ha^{-1} \, yr^{-1}$, the smaller figure deriving from Denmark with a relatively short growing season (Olsen, 1970; Milbank, 1974).

In cycads so called "coralloid roots" occur. These are a kind of root nodules infected with blue-green bacteria. Nitrogen fixation has been demonstrated in cycads and also in the tropical angiosperm, *Gunnera*, a herb which grows *Nostoc*, presumably heterotrophically, in glands in the stem and rhizome (Millbank, 1974; Nutman, 1976).

A general survey of blue-greens as symbiotic nitrogen fixers shows that in contrast to the root-nodule-forming Eubacteria they show practically no morphological differences relative to their free-living counterparts. Physiologically, however, they seem to have a higher rate of N_2-fixation and a lower growth rate than free-living forms and they release a large part of the assimilated nitrogen extracellularly to the benefit of their hosts (Millbank, 1974).

In aquatic environments some free-living, N_2-fixing blue-greens live in close contact with macrophytes (Capone and Taylor, 1977; Capone et al., 1977).

8.2.4. Symbiotic N_2-Fixation in Animals

Various reports on dinitrogen fixation by bacteria living in the intestinal tracts of animals (including ruminants and man) have appeared in the literature. Several anaerobes (e.g., *Clostridium*), known to have the capability to fix N_2, are found in the intestinal tract of animals. It is, however, very questionable whether any quantititative significance should be attributed to them. More recently, however, nitrogen fixation, carried out by *Citrobacter freundii* and *Enterobacter agglomerans*, has been shown to be important in the hind gut of termites (Breznak et al., 1973; French et al., 1976; Potrikus and Breznak, 1977).

8.3. PROKARYOTIC ENDOSYMBIONTS OF EUKARYOTE CELLS

During the last decade, much evidence has demonstrated a discontinuity between prokaryotes and eukaryotes. In order to explain the complexity of the eukaryote cell relative to prokaryotes, a theory first proposed at the turn of the century and which explains the origin of eukaryote organelles as being derived from endosymbiotic bacteria, has recently been revived and extended. The central idea of this theory, usually referred to as the "serial endosymbiotic theory", is that the original eukaryote ancestor was a hypothetical prokaryote which had acquired the ability of phagocytosis and which used only glycolysis as an energy yielding process. It then acquired aerobic endosymbiotic bacteria which evolved into mitochondria. Some of these now aerobic "proto-eukaryotes", then acquired endosymbiotic cyanobacteria which became the ancestors of plastids. Finally, ectosymbiotic, spirochaete-like bacteria are supposed to have been the origin of flagella and kinetoplasts. The theory, and the partially convincing evidence in its favor, is discussed in detail in, e.g., Cohen (1970), Margulis, (1970, 1971, 1975), Schenpf and Brown (1971), Hall (1973), and Taylor (1974). A number of observations and facts, however, are not so easily reconciled with the theory of serial endosymbiosis as it is usually presented (see e.g., Raff and Mahler, 1972; Stanier, 1974; Uzzell and Spolsky, 1974).

We will not here review all the evidence which is in favor or at variance with this theory. We believe that the theory, if not in all details, then at least with respect to its basic idea, offers the most convincing account for the origin of eukaryotes.

There are two reasons for mentioning the theory in this chapter. First of all it stresses the evolutionary significance of element cycling and energy yielding processes as expressed by the ideas on the origin of the mitochondrion and the chloroplast. Thus the symbiosis between an aerobic and a fermenting organism will provide the former with substrates (pyruvate and reduced NAD) which are waste products for the latter which in return receives ATP. Similarly a symbiotic relationship between photosynthetic and chemo-heterotrophic organisms provides the former with catabolic end products (CO_2, NH_3) and in return it synthesizes reduced carbon for the heterotroph. The basis for the hypothetical symbiotic relationships leading to the origin of mitochondria and plastids is therefore not principally different from the "syntrophisms" between different bacteria of microbial communities such as those of sediments or in the rumen.

Another reason for mentioning the theory of serial endosymbiosis is that it is a nice framework for discussing extant cases of prokaryotes which are endosymbionts of eukaryote cells; in fact the theory has led to an increased interest in this category of symbiosis during the last decade.

There are many examples of intracellular photosynthetic symbionts of heterotrophic eukaryotes. By far the greatest number of these are based on unicellular, eukaryotic algae (dinoflagellates, diatoms, etc.). In some cases it has turned out that the "symbionts" are only undigested chloroplasts deriving from phagocytosized algae and which are kept functional "*in vitro*" by the predator for some time (Taylor, 1970, 1973). *Endocyanella*, viz., intracellular cyanobacteria, are of course of special evolutionary interest but only relatively few cases have been established. *Geosiphon pyriforme* has been shown to be a symbiotic relationship between a phycomycete and a *Nostoc* species. The consortium which is capable of autotrophic growth, can be separated and both components can be grown independently in pure cultures if provided with the necessary nutrients. The endosymbiotic *Nostoc* cells do not seem to differ in morphological respects from the typical free-living blue-greens. Two other cases of endocyanosis, viz., *Cyanophora paradoxa* and *Glaucocystis nostochinearum* differ from that described above. Thus the systematic positions of the two hosts as well as the symbionts are enigmatic; *Cyanophora* is a kind of flagellate and *Glaucocystis* has been described as a chloroplast-free green algae. In spite of several attempts, it has not been possible to grow the components of the symbiosis separately. Ultrastructurally the endocyanelles differ significantly from typical blue-greens in lacking (in *Glaucocystis*) or having very reduced (in *Cyanophora*) cell wall; they are also abberant with respect to several other details. In some ways these two organisms resemble a unicellular rhodophyte (the chloroplasts of which have many structural and biochemical features in common with cyanobacteria) rather than protozoa with endosymbiotic blue-greens. *Cyanophora* and *Glaucocystis* may represent the best extant model of the evolution of chloroplasts from endocyanelles; most likely, however, the evolution of these photosynthetic protists was independent of the other plastid-containing eukaryotes and of more recent origin. A few other protozoa are also known to harbor endocyanella (Taylor, 1970; Schnepf and Brown, 1971; and other references therein).

Many eukaryotes, in particular protozoa, harbor intracellular bacteria. The associations are often obligatory and circumstantial evidence sometimes suggests that they have some physiological significance, but far the most cases are incompletely understood. A very interesting example is constituted by the freshwater amoeba *Pelomyxa palustris*. This organism does not possess mitochondria but endosymbiotic bacteria have taken over their respiratory role (Andresen, *et al.*, 1968; Leiner *et al.*, 1968). The few examples of obligatory anaerobic eukaryotes, viz., protozoans living in the intestinal tracts of some animals and in anaerobic sediments, seem nearly invariably to be associated with endosymbiotic (and often also ectosymbiotic) bacteria. It has been suggested that these bacteria utilize the end products of glycolysis which is probably the only energy yielding process open to these mitochondria-free protozoa (Fenchel *et al.*, 1977).

CHAPTER 9

Bacteria and Global Element Cycling

This last chapter is devoted to a discussion on the role of bacteria in the cycling of elements on a global scale. In Section 2.2, the prokaryote contribution to the evolution of the chemical environment of the surface of the earth was discussed; the present chapter concentrates on the extant biosphere and also summarizes some of the most important points of the previous chapters.

The global cycling of elements is often discussed in terms of compartment models with quantitative estimates of pool sizes and transfer rates such as shown in Figs. 46–48. Attempts to establish global balances of the important elements have been particularly popular during the last decade, since they are necessary to evaluate the quantitative effects of certain human activities such as the emission of CO_2 and SO_2 in conjunction with the combustion of fossil fuels. When presented in textbooks, the crudeness or even absence of quantitative data with respect to rates and reservoirs in the models is not often sufficiently emphasized. The constituents of the atmosphere and the hydrosphere are fairly accurately quantified but estimates of reservoirs in sediments and in living organisms and in particular of transfer rates are in many cases very inaccurate or only represent "educated guesses". In fact, even from a qualitative point of view, viz., the definition of pools and rates, such models may represent misleading simplifications. The fact that such presentations from different textbooks often show a close resemblance does not signify that the estimates are established truths, but rather the fidelity in the citation from previous compilations of this sort. Finally it must be emphasized that these models, at the best, represent a true description of the reality; the models do not explain the actual sizes of rates and pools. To this end, it is necessary to include considerations on chemical equilibria and kinetics as well as on biological and geological mechanisms. Some features of global element cycles may be explained but in view of the extreme complexity, formidable difficulties remain before a complete understanding will be achieved. Figures 46–48, which in the following discussion will be used as a framework, have been compiled from data in Hutchinson (1954), Bolin (1970), Cloud and Gibor (1970), Delwiche (1970) and Svenssen and Söder-

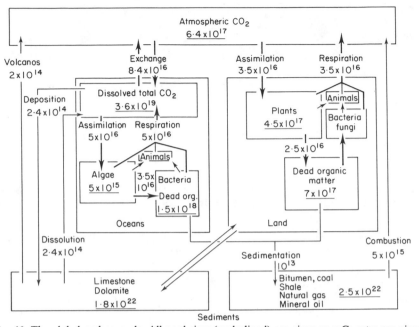

Fig. 46. The global carbon cycle. All pool sizes (underlined) are given as g C; rates are given as g C per year. The thick arrows indicate the quantitatively most important transfers.

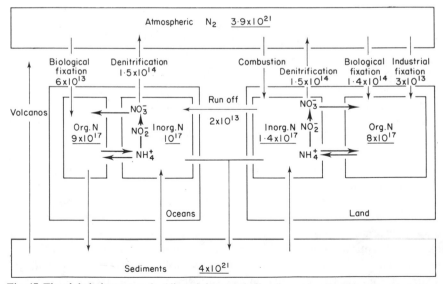

Fig. 47. The global nitrogen cycle. All pool sizes (underlined) are given as g N; rates are given as g N per year. The thick arrows indicate the quantitatively most important transfers.

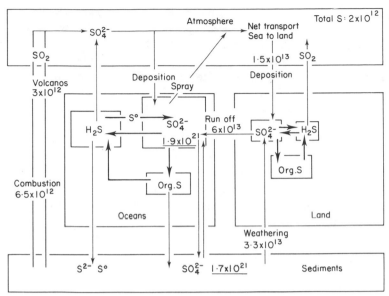

Fig. 48. The global sulfur cycle. All pool sizes (underlined) are given as gS; rates are given as gS per year. The thick arrows indicate the quantitatively most important transfers.

lund (1976); the last mentioned work has attempted a complete compilation of data relevant to the global cycling of nitrogen, sulfur and phosphorus. In our presentations we have not attempted to reconcile or balance the various inputs and outputs so that our biosphere may not seem to be in a completely steady state. They may, however, give a general idea about the order of magnitude with respect to some important reservoirs and pathways of transformation. We also hope that our presentations will not be quoted elsewhere as authoritative presentations of global element cycling.

Ever since the origin of life, its maintenance has been based on the creation of local chemical inequilibria by the electromagnetic radiation from the sun, i.e., the synthesis of organic molecules. This may have originally happened abiologically as discussed in Chapter 2, but for at least the last 3×10^9 years this has taken place in photosynthetic organisms. The potential chemical energy of the organic molecules is then the energetic basis of life. The energetic potential of the organic molecules increased as the environment became more oxidizing during the evolution of the biosphere. The tendency of the surface of the earth to become more oxidized was in part due to the abiotic photolysis of H_2O, CH_4 and NH_3 in the upper atmosphere, but it is first of all due to the advent of oxygenic photosynthetis. The main characteristic of the present biosphere is, therefore, a reduction–oxidation inequilibrium between the oxidizing atmosphere and hydrosphere on the one hand and the

reducing environment constituted by living organisms, and by the more or
less temporarily accumulated dead organic matter (and in some environments reducing inorganic molecules, e.g., H_2S, NH_3, CH_4, which derive from
the anaerobic mineralization of organic matter) on the other. This inequilibrium represents a dynamic state which can be described in terms of the
amount and turnover of O_2. There is about $1·2 \times 10^{21}$ g of free oxygen on the
surface of the earth, of which only about 1% is dissolved in the oceans, the
remaining constitutes about 21% of the atmosphere. The maintenance of
this oxygen is believed to be due to oxygenic photosynthesis and to a smaller
extent due to the photolysis of H_2O in the upper atmosphere. The turnover
time of atmospheric oxygen due to these O_2-producing processes and to
respiration and abiological oxidations is estimated to be about 2000 years.
If the steady state pool size of oxygen is the result of oxygenic photosynthesis, it presupposes a corresponding pool of reduced carbon. If only the
carbon of living biomass and of dead organic matter of ecosystems (perhaps
5×10^{17} and 2×10^{18} g, respectively) is considered, this would only account
for about $0·5\%$ of the free oxygen actually present. However, it is also necessary to include the fossil reduced carbon (bitumen, natural gas, mineral oil,
etc.) which is now believed to constitute $2·5 \times 10^{22}$ g C. This corresponds to
more than $6·7 \times 10^{22}$ g O_2 or more than 50 times the amount actually
present. The "missing" oxygen is, in part, found in the sulfate of the oceans
and the sediments; this alone represents about 7×10^{21} g O_2. The remaining
has, during geological time, been consumed by the oxidation of exposed
minerals, in particular the oxidation of ferrous to ferric iron. The available
data are, therefore, in accordance with the idea that the oxygen of the atmosphere is of photosynthetic origin, but as there are so many unknown
factors it is difficult to make a precise stoichiometric balance.

   Another way to give a general picture of the main properties of the biosphere is to consider water. This essential compound is, in the present biosphere, the main source of reducing power for photosynthesis. During photosynthesis the oxygen of the water is released; the water is re-formed during
respiratory processes. The turnover time of water, viz., the time taken before
an amount of water corresponding to that present on the surface of the earth
is split or recombined through biological processes is of the order of 2×10^6
years.

   In order to discuss specifically the role of bacteria in the transformation of
matter on a global scale, it is profitable to consider the cycling of the three
elements, C, N, and S. All three change valence during their cycling. There
are two reasons for this. They are all essential components of organisms, in
which they occur in a more reduced state than when in equilibrium with an
oxidized atmosphere. They, therefore, undergo a reduction in assimilatory
processes. Secondly, in a reduced state, they may all serve as electron donors

in the respiration of some organisms, whereas others can use these elements in an oxidized state as electron acceptors for the oxidation of reduced carbon in anaerobic respiration. Thus, these elements also change valence in dissimilatory, energy yielding processes. These generally have a lower energetic potential than the aerobic respiration of organic substrates and they are exclusively used by certain prokaryotes. However, the processes are necessary for the completion of the three element cycles and, therefore, essential for the functioning of the biosphere. In the following sections we will discuss the global cycling of C, N, and S separately and discuss the role of bacteria in transforming these elements.

9.1.   THE GLOBAL CARBON CYCLE

A crude presentation of the carbon cycle is shown in Fig. 46. An important feature (which the carbon cycle has in common with the other, biologically important element cycles) is that there are a number of smaller compartments (the atmospheric CO_2, the living organisms, and dead, non-sediment organic material) with a very rapid turnover; and some very large reservoirs (carbonates and organic carbon in sediments) which only exchange matter with the other pools at a very low rate. The carbon cycles of the oceans and of terrestrial environments are relatively self-contained but there is a rapid exchange between atmospheric CO_2 and dissolved CO_2 in the oceans. The CO_2 of the atmosphere is therefore turned over about every 7 years (this has been measured fairly accurately as the decrease in $^{14}CO_2$ of the atmosphere in the years following the large nuclear bomb tests around 1960). The partial pressure of atmospheric CO_2 is, therefore, largely regulated by the bicarbonate–carbonate system of the oceans. Thus, an increase in atmospheric CO_2 eventually leads to an increased deposition of calcite and dolomite, and a decrease leads to the dissolution of these minerals. Since the pH of the oceans, over a larger time scale, is buffered by silicate minerals it is likely that the partial pressure of the CO_2 of the atmosphere has been kept within relatively narrow limits (i.e., between 2×10^{-4} and 1.3×10^{-3} over a relatively long geological period. At the present, pCO_2 is about 3.2×10^{-4} but it tends currently to increase somewhat since the combustion of fossil fuels has increased faster than the rate at which CO_2 is absorbed in the sea (Bolin, 1970; Stumm and Morgan, 1970). This purely abiotic regulation of the pCO_2 may be an important regulator of photosynthetic activity in the biosphere.

As already discussed in Section 2.2, prokaryotic, anoxygenic and oxygenic, photosynthesis was responsible for all photoassimilation of CO_2 until late Precambrian times ($\sim 10^9$ years ago) and the cyanobacteria were probably responsible alone for the initial rise of O_2 content of the atmosphere. In the present biosphere (and probably at least throughout the whole Phanerozoic period) the photosynthetic activity of prokaryotes is quantitatively trivial. Anoxygenic photosynthesis is limited by the restricted co-occurrence of light and suitable elctron donors, and the blue-greens seem to be competitively inferior to eukaryotic photosynthesizers under most conditions. The latter (algae in the seas and vascular plants in terrestrial environments) are therefore totally dominating with respect to the assimilatory CO_2 reduction. Due to their dominating role, the photosynthetic plants also play a relatively large role in terms of returning some of the assimilated carbon as CO_2 to the atmosphere in respiration.

While prokaryotes play only a minor role with respect to reductive assimilation of CO_2, the opposite is true with respect to mineralization of organic matter. This was the theme of Chapter 3. There it was shown that their importance in breaking down organic matter is due to a number of special features of bacteria. These include the ability to hydrolyze whole classes of organic compounds (structural carbohydrates and other plant polymers and hydrocarbons) unavailable to eukaryotic heterotrophs (with the exception of some fungi under aerobic conditions), their ability to mineralize organics under anaerobic conditions and to utilize the reduced end-products of anaerobic metabolism under aerobic conditions, and finally their efficiency in utilizing dilute, dissolved substrates. There exists no estimate of the bacterial share in the CO_2 production of the biosphere but it may well exceed 50% of the total production rate. The mineralization of organic matter is the single most important quantitative role of bacteria.

A number of other bacterial contributions to the carbon cycle are of importance in this context, in particular with respect to fermentation processes. These have not been included in Fig. 46 and the quantitative importance of such processes on a global scale has not been extensively investigated. In particular methanogenesis could be mentioned in this context. The bacterial activities which influence the processes which eventually lead to the accumulation of fossil fuels must be of large biogeochemical significance although they are still quite incompletely understood.

9.2.   THE GLOBAL NITROGEN CYCLE

The global nitrogen cycle is shown schematically in Fig. 47. Among other features it differs from the carbon cycle in the presence of a very large atmospheric reservoir which contains nearly half of all nitrogen on the surface of the earth. Due to biological and abiological processes, N_2 is removed from the atmosphere and returned; the turnover time has been estimated to more than 10^7 years. The transfer rates to and from the atmospheric and sedimentary reservoirs of N are very small compared to the rapid exchange between dissolved inorganic and organic N-species (in particular NH_4^+, NO_2^-, and NO_3^-, and e.g., amino acids) and living organisms (Chapter 5).

The transformation of N-compounds carried out by eukaryotes, viz., the mineralization of organic N to NH_4^+ and the assimilatory reduction of NO_3^- are also carried out by bacteria; their quantitative role in this respect must more or less parallel their share in the carbon cycle. In addition, some transformations of N-compounds are unique to certain prokaryotes. These processes, which are treated in detail in Chapter 5, comprise denitrification, nitrification and nitrogen fixation.

The dinitrogen of the atmosphere may have originated through juvenile outgassing or through photolysis of NH_3 and it must have been a relatively conservative atmospheric constituent prior to the presence of oxygen above trace levels and/or the evolution of nitrogen fixation; these two events may be of comparable age (see Section 2.2). As discussed in Section 2.2.4, the process:

$$N_2 + 5/2O_2 + H_2O \rightarrow 2HNO_3$$

is exergonic and in spite of the high activation energy of the dinitrogen molecule it proceeds at a slow rate, e.g., in connection with electrical discharges in the atmosphere. The only known process of importance which maintains the nitrogen level of the atmosphere is microbial denitrification. Unfortunately, it is not yet possible to estimate the rate of denitrification on a global scale, it is generally believed to balance the loss of dinitrogen from the atmosphere (the quantitative estimates of which are also quite imprecise). A substantial amount of N_2O may be produced together with N_2 during denitrification. The fate of this in the atmosphere is not quite clear; in spite of its relatively high rate of production it occurs in very small concentrations but no photolytic process in the troposphere is known, which can explain its disappearance. It is believed that it is somehow decomposed into N_2, O_2 and nitrogen oxides. It may also be transformed biologically (for references see Svensson and Söderlund, 1976). Denitrification is also an important factor in regulating the availability of NO_3^-, which is usually totally depleted under anoxic conditions.

The opposite process, i.e., the transformation of dinitrogen into NO_3^-, may proceed directly abiotically (during atmospheric electrical discharges and in, e.g., combustion engines). Biologically, the process takes place indirectly. In nitrogen fixation the N_2 molecule is first reduced to NH_3, it is then subsequently ozidixed to NO_2^- and then to NO_3^- by nitrifying bacteria. The quantity of biologically fixed dinitrogen, on a global scale, is not known very precisely; the intensification in the study of this process and improved techniques have recently tended to increase estimates which currently are of the order of 2×10^{14} g N per year. The annual amount of industrially fixed N_2 which of course is much better known, constitutes a significant share of the total global N_2-fixation, perhaps 10–20%, and is, therefore, an example of human interference with global element cycling.

From the above it is clear that the nitrogen cycle of the biosphere is completed only through bacterially mediated processes which are of paramount importance for the maintenance of the distribution of nitrogen species in the atmosphere, the hydrosphere and in soils. It has not yet, however, been possible to propose a model which predicts the actual quantitative occurrence of nitrogen compounds and transfer rates.

As an example of mechanisms which may be important and which demonstrate how the nitrogen cycle is coupled to other element cycles, the stoichiometry of N and P in seawater may be mentioned. Marine planktonic algae have the empirical composition:

$$[C_{106}H_{263}O_{110}N_{16}P_1 + \text{trace elements}],$$

viz., a N/P ratio of 16. In seawater, N (as NO_3^-) and P (as HPO_4^{2-}) are also (with some local deviations) found in the ratio 16:1. Thus the two elements are depleted simultaneously as a function of algal growth. Different explanations are possible for this relationship. One possibility would be that the composition of algal biomass has evolved so as to adapt to the relative occurrences of the two elements. Alternatively and perhaps more likely, the controlling factor is phosphorus. The availability of phosphate is determined by the precipitation and dissolution of insoluble phosphates (Section 7.2). It is, therefore, possible that the level of NO_3^- is adjusted to the availability of P through denitrification, which is increased in the sediments when nitrate is in excess, and by nitrogen-fixation, which is favored if N is depleted prior to P (Stumm and Morgan, 1970).

9.3. THE GLOBAL SULFUR CYCLE

The last element to be considered here is sulfur. It is mainly found as sulfate in the oceans and in sedimentary rocks which also contain reduced and elemental sulfur. Atmospheric sulfur occurs only in low concentrations and is irregularly distributed; it occurs mainly in the form of SO_2 and SO_4^{2-} (in aerosols) and in trace amounts as H_2S, methyl mercaptan and methyl sulfides. The sulfur of the atmosphere is turned over rapidly since it is deposited with rain or as dry particles. Atmospheric sulfur is important since it constitutes the only transport from the sea to the land in addition to the weathering and erosion of S-containing minerals, and S is continuously lost as SO_4^{2-} to the sea in run-off with rivers. Plants and bacteria perform reductive assimilation of SO_4^{2-}. Unique to certain bacteria is dissimilatory sulfate reduction and chemoautotrophic and photoautotrophic oxidation of sulfide to elemental sulfur and sulfate. Sulfide may also be oxidized abiotically to $S°$, $S_2O_3^{2-}$ and SO_4^{2-}. Some organisms also liberate volatile sulfur compounds (mercaptans, methyl sulfides). Volcanoes and the combustion of fossil fuels add SO_2 to the atmosphere where it is oxidized to SO_4^{2-}. Sulfate is also added to the atmosphere in the form of seawater spray.

The microbial sulfur cycle, which is discussed in Chapter 6, is relatively well understood. It is now known that sulfate reduction is a quantitatively very important process in marine sediments where perhaps 50% of the carbon mineralization is catalyzed by this process. The underlying reason for this is first of all that sulfate constitutes an enormous capacity as electron acceptor in the seas whereas oxygen with its low solubility is rapidly depleted in stagnant water and in sediments. Thus, the total dissolved oxygen in the oceans constitutes about $4·8 \times 10^{19}$ oxidation equivalents compared to $1·5 \times 10^{22}$ or about 300 times more for sulfate.

The most important, and still unsatisfactorily answered question concerning global sulfur cycling is the biological contribution, in particular sulfate reduction, to the sulfur contents of the atmosphere. It is important since this information is needed in order to assess the global importance of SO_2 emission through the burning of fossil fuels (the local effect is well known for the erosion of limestone monuments in cities and the acidification of very oligotrophic lakes in the proximity of sources of air pollution). Of the enormous amounts of sulfide formed annually in sediments, swamps and soils, only a very small fraction can be expected to escape to the atmosphere; by far the most is rapidly reoxidized by bacteria or spontaneously, in close proximity to the site of formation or is accumulated as metallic sulfides. In oxygenated water, HS^- has a half life of the order of $0·5$ h so if only $0·5–1$ m oxidated water covers a sediment H_2S will not escape. Still, very shallow estuaries, tidal flats and salt marshes do emit large amounts of H_2S to the atmosphere, in particular during the night when photosynthetic bacteria are inactive and the pH of the water decreases as has been quantified by Hansen et al. (1978). There is not, however, any way to extrapolate such measurements to a global scale, due to the extreme patchiness of this type of environment. Also, the H_2S released just above the sediments may mainly have a local importance with only small amounts being mixed in the whole atmosphere, rather than being deposited or reabsorbed in the immediate surroundings of the source. Similar difficulties arise for evaluating the importance of volatile, organic S-compounds. The isotope composition of atmospheric S suggests a significant amount being of biological origin. However, due to the heterogeneous distribution of atmospheric sulfur and the unknown amounts deriving from seawater spray, the stable isotope fractionation cannot be used for a global quantification of biologically derived, atmospheric sulfur.

The Calculation of Free Energy Yields of Biochemical Reactions

The $\Delta G_o'$ of a reaction may be calculated from the standard free energies of formation, ΔG_f° (25°C), of the reactants and the products. This usually involves obtaining a small difference from large numbers. The result is that proportional small errors in the free energies of formation can lead to relatively large proportional errors in the $\Delta G_o'$ values:

$$\Delta G^\circ = \text{(sum of } \Delta G_f^\circ \text{ of products)} - \text{(sum of } \Delta G_f^\circ \text{ of reactants)}.$$

For example, in the reaction involving the oxidation of acetate by sulfate:

$$CH_3COO^- + SO_4^{2-} + 2H^+ \rightleftharpoons 2CO_2 + 2H_2O + HS^-$$

$$\Delta G^\circ = (2 \times -92.26 + 2 \times -56.687 + 2.88) - (-88.29 - 177.97)$$

$$= -28.754 \text{ kcal mol}^{-1}.$$

The $\Delta G_o'$ is defined as the free energy change at pH 7 and it is therefore necessary to take the hydrogen ion concentration into account.

$$\Delta G_o' = \Delta G^\circ + 2.303 \; RT \log \frac{[CO_2]^2 \; [H_2O]^2 \; [HS^-]}{[CH_3COO^-] \; [SO_4^{2-}] \; [H^+]^2};$$

since all the reactants and products are at unit activity:

$$\Delta G_o' = \Delta G^\circ + 2.303 \; RT \log 1/[H^+]^2, \text{ and } \log 1/[H^+] = pH.$$

Therefore $\Delta G_o' = \Delta G^\circ + 2.303 \times 1.987 \times 10^{-3} \times 298 \times 2 \times 7$

$$= -9.663 \text{ kcal mol}^{-1}.$$

Similarly for:

$$NH_4^+ + 3/2O_2 \rightleftharpoons NO_2^- + 2H^+ + H_2O;$$

$$\Delta G^\circ = -46.62 \text{ kcal mol}^{-1}, \text{ and}$$

$$\Delta G_o' = -65.71 \text{ kcal mol}^{-1}.$$

182

The ΔG_f° of a number of molecules discussed in the text are listed below. These values, which are taken from Thauer *et al.* (1977) may be converted to $kJ\,mol^{-1}$ by multiplication with 4·184.

Substance	State	ΔG_f° $(kcal\,mol^{-1})$
H_2O	liq	56·687
CO_2	aq	92·26
CH_4	g	12·13
methanol	aq	41·92
ethanol	aq	43·44
formaldehyde	aq	31·20
formate	aq	83·9
acetate	aq	88·29
propionate	aq	86·3
butyrate	aq	84·28
fumarate	aq	144·41
succinate	aq	164·97
lactate	aq	123·6
α-D-glucose	aq	219·22
L-alanine	aq	88·8
glycine	aq	88·618
NH_4^+	aq	18·97
NO	g	$-20·69$
NO_2^-	aq	8·9
NO_3^-	aq	26·61
N_2O	g	$-24·90$
SH^-	aq	$-2·88$
SO_3^{2-}	aq	116·3
SO_4^{2-}	aq	177·97
$S_2O_3^{2-}$	aq	122·7
Fe^{2+}	aq	18·85
Fe^{3+}	aq	1·1

The Classification of Prokaryotes

The classification of organisms serves two purposes. It is a method for bringing order to the great and chaotic number of organisms, thus aiding identification and providing means to characterize groups of forms which share certain properties. Secondly it is an attempt to trace the natural relationships, viz., the phylogeny and evolution of different groups of organisms. The taxonomy of prokaryotes differs in several respects from that of the eukaryotes. Thus the species concept used for eukaryotic organisms is not easily applied to organisms where no reproductive units, viz., populations of which the individuals share a common gene pool, exist. In a sense every bacterial clone constitutes its own species. In spite of this, fortunately bacteria do not show a total continuum of characters; in nature they usually appear to form discrete groups within which isolated clones share all or nearly all properties. Such groups are designated species but this is an arbitrary procedure. Prokaryotic taxonomy is also difficult due to the fact that these organisms show relatively little morphological diversity and those features which exist (cell shape, colony formation, type of motility, etc.) often do not correlate with biochemical features (type of metabolism, cell wall chemistry, etc.) which are usually considered to trace phylogeny more correctly. In some cases this is probably not true; for example *Beggiatoa* is often classified together with other chemolithotrophic bacteria oxidizing reduced sulfur, but most probably it is derived from the cyanobacterium *Oscillatoria* to which it shows close resemblance. Finally it must be remembered that the prokaryotes have diverged and evolved for at least 3×10^9 years which is much longer than eukaryotes; already for this reason the phylogeny of the former is very difficult to reconstruct.

More recent methods have given some promise of an improved understanding of the true relationships at the generic and specific level (e.g., DNA-base pairing) as well as on a macroevolutionary scale (the $G + C$ ratios of ratios of DNA, amino acid sequences of homologous proteins). An increased theoretical understanding of the evolution of biochemical pathways in metabolism may also contribute to that end (see also Section 2.2). However, bacterial classification is still mainly a purely empirical enterprise and it serves to bring some kind of order to the many forms described and facilitates identification.

The prokaryotes are sometimes divided into four major groups: Eubacteria, Cyanobacteria (= blue-green algae), Myxobacteria and Spirochaeta. The main characteristics of these groups are shown in Table XXIII. Below, we give for each of these groups the main sub-groups and within each of these we will briefly mention each of the about 100 genera of bacteria we have used as examples in the book. The purpose of this list is first of all to give the non-microbiologist the possibility of a first orientation in the prokaryote kingdom and to be able to place the forms mentioned in the book within some framework.

The classification scheme used below is mainly based on Stanier *et al.* (1971) to which the reader is also referred for further discussion on the problems of bacterial classification and for references.

TABLE XXIII. The Major Groups of Prokaryotes.

	Eubacteria	Cyanobacteria	Spirochaeta	Myxobacteria
Cell wall	thick, rigid	thick, rigid	thin, flexible	thin, flexible
Motility	flagella or −	gliding or −	contractile	gliding
Resting stages	endospores	akinetes		microcysts
Metabolism:				
oxygenic				
photosynthesis	−	+	−	−
anoxygenic				
photosynthesis	+	+	−	−
chemolithotrophy	+	+	−	−
chemoheterotrophy	+	+	+	+

EUBACTERIA

1. Gram-Negative Bacteria

1. Photosynthetic Bacteria

Eubacteria with anoxygenic phototrophy. A morphologically very diverse group of mainly obligate anaerobes. They require reducing conditions and light at the same time and are mainly found in stratified water bodies and in estuarine sediments. Three groups are usually recognized. The *green sulfur-bacteria* contain mainly bacteriochlorophyll *b* or *c* but always with small amounts of bacteriochlorophyll *a*. They are strict anaerobes and utilize H_2S or H_2 as electron donors in photosynthesis and they do not store intracellular sulfur: *Chlorobium*. "*Chlorochromatium*" is a symbiotic association of a heterotrophic bacterium covered by *Chlorobium* cells. The *purple sulfurbacteria* contain bacteriochlorophyll *a* or *d*, they utilize H_2S or $S°$ as electron donors but they do not tolerate very high sulfide concentrations. They mostly store $S°$ in the cells. They often show mass occurrences which are easily detectable due to their red or purple carotenoid pigments: *Chromatium, Ectothiorhodospira, Thiocapsa, Thiocystis, Thiopedia*. The *purple non-sulfur bacteria* resemble the purple sulfurbacteria with respect to pigments. They utilize reduced organics or H_2 as electron donors. Some are tolerant to oxygen and can grow heterotrophically in the dark: *Rhodopseudomonas, Rhodospirillum. Chloroflexis* is a filamentous photoheterotroph with pigments resembling those of the green bacteria; it is found in hot springs. All photosynthetic bacteria investigated can fix N_2.

2. Enteric Bacteria

Small flagellated or immobile rods which grow aerobically on a variety of organic compounds or anaerobically by fermenting carbohydrates in mixed acid fermentation with or without the production of H_2 and CO_2; some are denitrifiers. Many are found as commensals or pathogens in the intestinal tract of vertebrates: *Escherichia*; others are often or exclusively, found in soil and water: *Klebsiella, Enterobacter, Citrobacter, Aerobacter*, all fix N_2 when growing anaerobically; *Serratia, Proteus, Vibrio*.

3. The Chemolithotrophic Bacteria

(a) The *hydrogen bacteria* (*Hydrogenomonas*) oxidize H_2; they are all facultative chemoautotrophs and may be classified together with *Pseudomonas*.

(b) The *nitrifying bacteria* are mostly obligatory chemolithotrophs. *Nitrosomonas, Nitrosocystis, Nitrosobolus*, and *Nitrosospira* oxidize NH_3 to NO_2^-; *Nitrobacter, Nitrococcus*, and *Nitrospina* oxidize NO_2^- to NO_3^-.

(c) The *white sulfurbacteria* oxidize reduced sulfur compounds to $S°$ or SO_4^{2-}. *Thiobacillus* comprise forms which oxidize sulfide, elemental sulfur or thiosulfate to sulfate; *T. ferrooxidans* can also grow autotrophically by oxidizing ferrous iron in acid environments. *Sulfolobus, Thiovolum, Thiophysa*, and *Achromatium* are large bacteria which store elemental sulfur in the cells; chemolithotrophy has not been proven in all cases.

(d) *Bacteria oxidizing C-1 compounds. Methanomonas* oxidizes CH_4 and other C-1 compounds. A number of chemoheterotrophic bacteria (e.g., *Pseudomonas*) can also grow on C-1 compounds.

4. The Aerobic Chemoheterotrophs

Pseudomonas spp. utilize a variety of organic molecules; many of the species are capable of nitrate respiration and they are common in soils and water. The related halophilic bacteria (e.g., *Halobacterium*) have an absolute requirement for high NaCl concentrations. Other genera within the group are *Alginobacter*, *Agarbacterium*, *Alginomonas*, *Flavobacterium*, *Achromobacter*, and *Alcaligenes*. The acid-tolerant acetic acid bacteria which oxidize ethanol to acetate also belong here. *Spirillum* spp. are morphologically distinct; they are common in decomposing material in soils and water. *Bdellovibrio* is a small "parasite" or "predator" of other Gram-negative bacteria. *Azotobacter* is an aerobic, free-living nitrogen fixer; its large cells are found in soils and water. *Rhizobium* spp. fix N_2 in symbiosis with legume roots.

5. Aggregate-Forming Bacteria

Some of the Gram-negative, aerobic chemoheterotrophs form colonies. *Zoogloea* and *Lamprocystis* form irregular, gelatinous colonies and sheaths of regularly arranged cells respectively. *Siderocapsa*, *Sphaerotilus*, *Crenothrix*, *Leptothrix*, and *Cladothrix* form colonies of longitudinally arranged cells within cylindrical sheaths which are often encrusted with iron and manganese oxides. All these forms are characteristic inhabitants of polluted waters and streams.

6. The Stalked Bacteria

These aerobic chemoheterotrophs form extensions of the cell wall which may serve as a holdfast. Cell fission can be described as a budding and the life cycle may be polymorphic. Many are known to grow in very dilute substrate concentrations and they have a low growth rate. *Caulobacter*, *Assticacaulis*, *Hyphomicrobium*, *Gallionella*, claimed to be a chemolithotrophic oxidizer of reduced iron and which form twisted filaments of iron hydroxide and *Metallogenium* which also mediates the oxidation of reduced iron and manganese, may not have any real relationship to the other bacteria in this group.

7. The Obligate Anaerobe Gram-Negative Bacteria

Desulfovibrio uses sulfate as the terminal electron acceptor for oxidizing certain organic substrates (e.g., lactate) and H_2 thus forming sulfide. They are abundant in all anaerobic environments containing sulfate. *Desulfuromonas* can oxidize acetate with elemental sulfur. Other representatives of the group are best known from the digestive system of ruminants; they perform different fermentative processes and many can hydrolyze structural carbohydrates: *Bacteroides*, *Butyrivibrio*, *Selenomonas*, *Veillonella*; see also Section 8.1.1.

2. Gram-Positive Bacteria

1. Aerobic Spore Formers

The genus *Bacillus* comprises obligatory aerobes or facultative anaerobes;

one form is a denitrifier. They can utilize a number of organic substrates and are common in soils.

2. Anaerobic Spore formers

The genus *Clostridium* constitutes a group of obligate anaerobes; a large variety of fermentation processes are represented within the genus and some species can hydrolyze structural carbohydrates and others are proteolytic; the ability to fix nitrogen is widespread. There are many species in soils and other anaerobic environments. Species which perform anaerobic respiration by dissimilatory sulfate reduction are often assigned to the genus *Desulfotomaculum*.

3. Lactic Acid Bacteria

Not spore forming; mainly ferment carbohydrates into lactic acid; they are oxygen tolerant. Their rapid growth and tolerance to their own metabolic product, lactic acid, are adaptations to life in high concentrations of carbohydrates such as found in milk, decaying fruits, etc.: *Lactobacillus, Streptococcus.*

4. The Methane Bacteria

Strict anaerobes which use CO_2 as an electron acceptor in anaerobic respiration thus producing CH_4. In anoxic sediments, anaerobic sewage digestors and in the rumen: *Methanobacterium, Methanosarcina, Methanococcus*, with respect to "*Methanobacillus omelianskii*", see Section 8.1.

5. Other Gram-Positive Cocci

Sarcina is an oxygen tolerant fermenter of soluble carbohydrates, it can synthesize cellulose. *Micrococcus* contains some obligate anaerobe forms; *Ruminococcus, Paracoccus.*

6. Coryneform Bacteria

These are immobile bacteria; they may be aerobes, facultative anaerobes or oxygen tolerant fermentors. *Corynebacterium* ferments sugars to propionic acid. *Propionibacterium* performs propionic acid fermentation of carbohydrates or of lactate in the rumen; one species is a denitrifier. *Cellulomonas* lives in soils and hydrolyzes cellulose. *Arthrobacter*, also a soil inhabitant, can attack halogenated aromatic compounds. *Brevibacterium* is an aerobe living in milk.

7. The Proactinomycetes

Aerobic bacteria which reproduce by fragmentation of mycelial cells. Slowly growing organisms which are common in soils. They can utilize hydrocarbons and other large organic molecules; *Nocardia Mycobacterium.*

8. The True Actinomycetes

Mycelial organisms reproducing through the formation of conidospores. Very important, aerobic bacteria in soils. They can utilize various structural compounds: *Streptomyces*.

CYANOBACTERIA

1. Unicellular or in colonies imbedded in mucus, immobile

Eucapsis, Microcystis.

2. Filamentous forms or discrete cells within a common cell wall

A. *Oscillatoria*-types, no heterocysts: *Oscillatoria*.
B. *Nostoc*-types with heterocysts: *Anabena, Nostoc*.
C. *Rivularia*-types, with heterocysts usually attached to a substratum: *Calothrix*.

3. Non-photosynthetic forms

A. H_2S oxidizing, presumably chemolithotrophic forms: *Beggiatoa, Thiothrix, Thioploca*.
B. Chemoheterotrophic forms: *Vitreoscilla*.

The different non-photosynthetic forms can, on a morphological basis, be classified together with the different photosynthetic forms; thus *Beggiatoa* is very close to *Oscillatoria* and *Thiothrix* is closely related to *Calothrix*.

MYXOBACTERIA

1. The Cytophagas

Aerobic, flexible, gliding cells. Can break down cellulose, chitin and other structural carbohydrates: *Cytophaga* lives in soils and water.

2. The fruiting Myxobacteria

May not have any phylogenetic relationship to the cytophagas. Morphology and life cycle show superficial resemblance to the (eukaryotic) slime moulds. Cells aggregate to form complex fruiting bodies. Common in soils: *Sorangium* is a cellulose decomposer.

SPIROCHAETA

Common chemoheterotrophs in soils and water; no genera belonging to this group of prokaryotes have been mentioned in this book.

References

Abeliovich, A. and Shilo, M. (1972). Photoxidative death in blue-green algae. *J. Bacteriol.* **111**, 682–689.

Ahearn, D. G. and Meyers, S. P., eds. (1973). "The Microbial Degradation of Oil Pollutants". Lousiana State Univ. Publ. **73-01**. Baton Rouge, La.

Alexander, M. (1961). "Introduction to Soil Microbiology". John Wiley & Sons, Inc., New York.

Alexander, M. (1965). Nitrification. *Agronomy* **10**, 307–343.

Ali, S. H. and Stokes, J. L. (1971). Stimulation of heterotrophic and autotrophic growth of *Sphaerotilus discophorus* by manganous irons. *Antonie van Leeuwenhoek* **37**, 519–528.

Allen, H. L. (1971). Dissolved organic carbon utilization in size-fractionated algal and bacterial communities. *Internat. Rev. Gesamt. Hydrobiol.* **56**, 731–749.

Allen, R. C., Gavish, E., Friedman, G. M. and Sanders, J. E. (1969). Aragonite-cemented sandstone from outer continental shelf off Delaware Bay: submarine lithification mechanisms yields product resembling beachrock. *J. Sediment. Petrol.* **39**, 136–149.

Anderson, J. H. (1965). Studies on the oxidation of ammonia by *Nitrosomonas*. *Biochem. J.* **95**, 688–698.

Anderson, J. M. (1977). Rates of denitrification of undisturbed sediment from six lakes as a function of nitrate concentration, oxygen and temperature. *Arch. Hydrobiol.* **80**, 147–159.

Andreesen, J. R., Schaupp, A., Neuranter, C., Brown, A. and Ljungdahl, L. G. (1973). Fermentation of glucose, fructose and xylose by *Clostridium thermoaceticum*: effects of metals on growth yield, enzymes and synthesis of acetate from CO_2. *J. Bacteriol.* **114**, 743–751.

Andresen, N., Chapman-Andresen, C. and Nilsson, J. R. (1968). The fine structure of *Pelomyxa palustris*. *C. R. trav. lab. Carlsberg* **36**, 285–317.

Ault, W. V. and Kulp, J. L. (1959). Isotopic geochemistry of sulfur. *Geochim. Cosmochim. Acta.* **16**, 201–235.

Baas-Becking, L. G. M. (1925). Studies on the sulfur bacteria. *Ann. Botany* **39**, 613–650.

Balderston, W. L. and Payne, W. J. (1976). Inhibition of methanogenesis in salt marsh sediments and whole-cell suspensions of methanogenic bacteria by nitrogen oxides. *Appl. Environ. Microbiol.* **32**, 264–269.

Balderston, W. L., Sherri, B. and Payne, W. J. (1976). Blockage by acetylene of nitrous oxide reduction in *Pseudomonas perfectomarinus*. *Appl. Environ. Microbiol.* **31**, 504–508.

Barghoorn, E. S. (1971). The oldest fossils. *Sci. Amer.* **224:5**, 30–42.

Barker, H. A. (1956). "Bacterial Fermentations", Wiley, New York.

Barnes, R. O., Bertine, K. K. and Goldber, E. D. (1975). N_2:Ar, nitrification and denitrification in southern California borderland sediments. *Limnol. Oceanog.* **20**, 962–970.

Barnes, R. O. and Goldberg, E. D. (1976). Methane production and consumption in anoxic marine sediments. *Geology* **4**, 297–300.

Baross, J. A., Hanus, F. J., Griffiths, R. P. and Morita, R. Y. (1975). Nature of incorporated ^{14}C-labelled material retained by sulfuric acid fixed bacteria in pure cultures and in natural aquatic populations. *J. Fish. Res. Board Can.* **32**, 1876–1879.

Barsdate, R. J., Fenchel, T. and Prentki, R. T. (1974). Phosphorus cycle of model ecosystems: significance for decomposer food chains and effect of bacterial grazers. *Oikos* **25**, 239–251.

Bauchop, T. and Elsden, S. R. (1960). The growth of microorganisms in relation to their energy supply. *J. Gen. Microbiol.* **23**, 457–469.

Baxter, R. M., Wood, R. B. and Prosser, M. V. (1973). The probable occurrence of hydroxylamine in the water of an Ethiopian lake. *Limnol. Oceanog.* **18**, 470–472.

Bella, D. A., Ramm, A. E. and Peterson, P. E. (1972). Effects of tidal flats on estuarine water quality. *J. Water Pollut. Control Fed.* **44**, 541–556.

Belly, R. T. and Brock, T. D. (1974). Ecology of iron-oxidizing bacteria in pyritic materials associated with coal. *J. Bact.* **117**, 726–732.

Bender, M. L., Fanning, K. A., Froelich, P. N., Heath, G. R. and Maynard, V. (1977). Interstitial nitrate profiles and oxidation of sedimentary organic matter in the eastern equatorial Atlantic. *Science* **198**, 605–609.

Berkner, L. V. and Marshall, L. C. (1967). The rise of oxygen in the earth's atmosphere with notes on the Martian atmosphere. *Adv. Geophys.* **12**, 309–331.

Berner, R. A. (1970). Sedimentary pyrite formation. *Am. J. Sci.* **268**, 1–23.

Berner, R. A. (1971). "Principles of Chemical Sedimentology". McGraw-Hill Book Co. New York.

Berner, R. A. (1974). Kinetic models for the early diagenesis of nitrogen, phosphorus, and silicon in anoxic marine sediments. *In* "The Sea" (E. D. Goldberg, ed.), **5**, 427–450. Interscience, New York.

Billen, G. (1975). Nitrification in the Scheldt estuary (Belgium and the Netherlands). *Estuarine Coastal Mar. Sci.* **3**, 79–89.

Blackburn, T. H. (1965). Nitrogen metabolism in the rumen. *In* "Physiology of Digestion in the Ruminant" (R. W. Dougherty, ed.) pp. 322–334. Butterworths, Washington.

Blackburn, T. H. (1979a). N/C ratios and rates of ammonia turnover in anoxic sediments. *In* "Microbial Degradation of Pollutants in Marine Environments" (A. W. Bourquin and P. H. Pritchard, eds.) (in press).

Blackburn, T. H. (1979b). A method for measuring rates of NH_4^+ turnover in anoxic marine sediments using a ^{15}N–NH_4^+ dilution technique. *Appl. Environ. Microbiol* **37**, (in press).

Blackburn, T. H. and Hungate, R. E. (1963). Succinic acid turnover and propionic production in the bovine rumen. *Appl. Microbiol.* **11**, 132–135.

Blackburn, T. H., Kleiber, P. and Fenchel, T. (1975). Photosynthetic sulfide oxidation in marine sediments. *Oikos* **26**, 103–108.

Bolin, B. (1970). The carbon cycle. *Sci. Amer.* **223:3**, 124–132.

Bollag, J. M. and Czlonkowski, S. T. (1973). Inhibition of methane formation in soil by various nitrogen containing compounds. *Soil Biol. Biochem.* **5**, 673–678.

Bond, G. (1963). The root nodules of non-leguminous angiosperms. *Symposia Soc. Gen. Microbiol.* **13**, 72–91.

Bond, G. (1974). Root-nodule symbiosis with actinomycete-like organisms. *In* "The Biology of Nitrogen Fixation" (A. Quispel, ed.) pp. 342–378. North Holland Publishing Company, Amsterdam.

Bowman, R. A. and Focht, D. D. (1974). The influence of glucose and nitrate upon denitrification rates in sandy soils. *Soil Biol. Biochem.* **6**, 297–301.

Boysen-Jensen, P. (1914). Studies concerning the organic matter of the sea bottom. *Rep. Dan. Biol. Sta.* **22**, 1–39.

Bremner, J. M. and Bundy, L. G. (1974). Inhibition of nitrification in soils by volatile sulfur compounds. *Soil Biol. Biochem.* **6**, 161–165.

Breznak, J. A., Brill, W. S., Mertins, J. W. and Coppel, H. C. (1973). Nitrogen fixation in termites. *Nature* **244**, 577–580.

Breznak, J. A. and Pankrantz, H. S. (1977). In situ morphology of gut microbiota of wood-eating termites (*Reticulitermes flavipes* (Kollar) and *Coptotermes formosanus* (Shiraki)). *Appl. Environ. Microbiol.* **33**, 406–426.

Brezonik, P. L. (1972). Nitrogen: sources and transformations in natural waters. *In* "Nutrients in Natural Waters" (H. E. Allen and J. R. Kramer, eds.) pp. 1–50. John Wiley & Sons, New York.

Brill, W. J. (1975). Regulation and genetics of bacterial nitrogen fixation. *Ann. Rev. Microbiol.* **29**, 109–130.

Brill, W. S. (1977). Biological nitrogen fixation. *Sci. Amer.* **236**:3, 68–81.

Brock, T. D. (1973a). Lower pH limit for the existence of blue-green algae: evolutionary and ecological implications. *Science* **175**, 480–483.

Brock, T. D. (1973b). Evolutionary and ecological aspects of the Cyanophytes. *In* "The Biology of Blue-Green Algae" (N. G. Carr and B. A. Whitton, eds.) pp. 487–500. Blackwell, Oxford.

Broda, E. (1975a). "The Evolution of the Bioenergetic Process". Pergamon Press, Oxford.

Broda, E. (1975b). The history of inorganic nitrogen in the biosphere. *J. Mol. Evol.* **7**, 87–100.

Broda, E. (1977). Two kinds of lithotrophs missing in nature. *Zeitschrift für Allg. Mikrobiol.* **17**, 491–493.

Broecker, W. S. (1970). A boundary condition on the evolution of atmospheric oxygen. *J. Geophys. Res.* **75**, 3553–3557.

Bromfield, S. M. (1954). Reduction of ferric compounds by soil bacteria. *J. Gen. Microbiol.* **11**, 1–6.

Brooks, M. A. (1963). Symbiosis and aposymbiosis in arthropods. *Symposia Soc. Gen. Microbiol.* **13**, 200–231.

Bryant, M. P. (1969). Ethanol and lactate fermentation by anaerobic bacteria in association with H_2-using methanogenic bacteria. *Abstract*, 158 *National Meeting, Am. Chem. Soc.*, No. 18.

Bryant, M. P. (1976). The microbiology of anaerobic degradation and methanogenesis with special reference to sewage. *In* "Microbial Energy Conversion". (H. G. Schlegel and J. Barnea, eds.), pp. 107–117. Erich Goltze KG, Gottingen.

Bryant, M. P., Varec, V. H., Frobish, R. A. and Isaacson, H. R. (1976). Biological potential of thermophilic methanogenesis from cattle wastes. *In* "Microbial Energy Conversion" (H. G. Schlegel and J. Barnea, eds.), pp. 347–359. Erich Goltze KG, Göttingen.

Bryant, M. P., Wolin, E. A., Wolin, M. J. and Wolfe, R. S. (1967). *Methanobacillus omelianskii*, a symbiotic association of two species of bacteria. *Arch. Microbiol.* **59**, 20–31.

Burges, A. and Raw, F., eds. (1967). "Soil Biology", Academic Press, London.

Burns, R. C. and Hardy, R. W. F., eds. (1975). "Nitrogen Fixation in Bacteria and Higher Plants". Springer-Verlag, Heidelberg.

Buswell, A. M., Shiota, T., Lawrence, N. and Meter, I. W. (1954). Laboratory studies on the kinetics of growth of *Nitrosomonas* with relation to the nitrification phase of the BOD test. *Appl. Microbiol.* **2**, 21–25.

Byrnes, B. H., Keeny, D. R. and Graetz, D. A. (1972). Release of ammonium-N from sediments to water. *Proc. 15 Conf. Great Lakes Res.* 1972, 249–254.

Campbell, N. E. R. and Aleem, M. I. H. (1965). The effect of 2-chloro-6-(trichloro-methyl)-pyridine on the chemoautotrophic metabolism of nitrifying bacteria. 2. Nitrite oxidation by *Nitrobacter*. *Antonie van Leeuwenhoek*, **31**, 124–136.

Campbell, N. E. R. and Lees, H. (1967). The nitrogen cycle. *In* "Soil Biochemistry" (A. D. McLaren and G. H. Peterson, eds.), pp. 194–215. Marcel Dekker, New York.

Campbell, L. L., Jr. and Williams, O. B. (1951). A study of chitin-decomposing micro-organisms of marine origin. *J. Gen. Microbiol.* **5**, 894–905.

Capone, D. G. and Taylor, B. F. (1977). Nitrogen fixation (acetylene reduction) in the phyllosphere of *Thalassia testudinum*. *Mar. Biol.* **40**, 19–28.

Capone, D. G., Taylor, D. L. and Taylor, B. F. (1977). Nitrogen fixation (acetylene reduction) associated with macro-algae in a coral reef community in the Bahamas. *Mar. Biol.* **40**, 29–32.

Cappenberg, T. E. (1974). Interrelations between sulfate-reducing and methane-producing bacteria in the bottom deposits of a freshwater lake. II. Inhibition experiments. *Antonie van Leeuwenhoek* **40**, 297–306.

Cappenberg, T. E. (1975). A study of mixed continuous cultures in sulfate-reducing and methane-producing bacteria. *Microbial Ecol.* **2**, 60–72.

Cappenberg, T. E. and Prins, R. A. (1974). Interrelations between sulfate-reducing and methane-producing bacteria in bottom deposits of a freshwater lake. III. Experiments with ^{14}C-labelled substrates. *Antonie van Leeuwenhoek* **40**, 457–469.

Carr, N. G. and Whitton, B. A., eds. (1973). "The Biology of the Blue-Green Algae". Blackwell, Oxford.

CAST. (1976). Effects of increased nitrogen fixation on stratospheric ozons. *Rep.* **53**. *Councils for Agricultural Science and Technology*, Iowa.

Castenholz, R. W. (1973). The possible photosynthetic use of sulfide by filamentous phototrophic bacteria of hot springs. *Limnol. Oceanog.* **18**, 863–876.

Chen, R. L., Keeney, D. R., Groetz, D. A. and Holding, J. A. (1972a). Denitrification and nitrate reduction in Wisconsin lake sediments. *J. Environ. Quality* **1**, 158–162.

Chen, R. L., Keeney, D. R. and Konrad, J. G. (1972b). Nitrification in sediments of selected Wisconsin lakes. *J. Environ. Quality.* **1**, 151–154.

Chen, K-Y. and Morris, J. C. (1972). Kinetics of oxidation of aqueous sulfide by O_2. *Environ. Sci. Technol.* **6**, 529–537.

Cho, C. M. (1971), Convective transport of ammonium with nitrification in soil. *Can. J. Soil Sci.* **51**, 339–350.

Clark, P. H. (1974). The evolution of enzymes for the utilization of novel substrates. *Symposia Soc. Gen. Microbiol.* **24**, 183–217.

Claypool, G. and Kaplan, I. R. (1974). The origin and distribution of methane in sediments. *In* "Natural Gases in Marine Sediments" (I. R. Kaplan, ed.), pp. 99–139. Plenum Press, New York and London.

Cloud, P. (1974). Rubey conference on crustal evolution. *Science* **183**, 878–881.

Cloud, P. (1976). Beginnings of biospheric evolution and their biochemical consequences. *Palaeobiol.* **2**, 351–387.

Cloud, P. and Gibor, A. (1970). The oxygen cycle. *Sci. Amer.* **223**:3, 111–123.

Codisponti, L. A. and Richards, F. A. (1976). An analysis of the horizontal regime of denitrification in the eastern tropical north Pacific. *Limnol Oceanogr.* **21**, 379–388.

Cohen, S. S. (1970). Are/were mitochondria and chloroplasts microorganisms? *Amer. Sci.* **58**, 281–289.

Cohen, Y., Jørgensen, B. B., Padan, E. and Shilo, M. (1975). Sulfide-dependent anoxygenic photosynthesis in the cyanobacterium *Oscillatoria limnetica. Nature* **257**, 489–492.

Connell, W. E. and Patrick, W. H., Jr. (1969). Reductions of sulfate to sulfide in waterlogged soil. *Soil Sci. Soc. Amer. Proc.* **33**, 711–715.

Cosgrove, D. J. (1977). Microbial transformations in the phosphorus cycle. *Adv. Microbiol Ecol.* **1**, 95–134.

Cowan, G. A. (1976). A natural fission reactor. *Sci. Amer.* **235:1**, 36–47.

Crawford, C. C., Hobbie, J. E. and Webb, K. L. (1974). The utilization of dissolved free amino acids by estuarine microorganisms. *Ecology* **55**, 551–563.

Crutzen, P. J. (1974). Estimation of possible variations in total ozone due to natural causes and human activities. *Ambio* **3**, 201–210.

Culver, D. A. and Brumskill, G. J. (1969). Fayetteville Green Lake, New York, V. Studies of primary production and zooplankton in a meromictic lake. *Limnol. Oceanogr.* **14**, 862–873.

Czeczuga, B. (1968). An attempt to determine primary production of green sulphur bacteria, *Chlorobium limnicola* Nads, (Chlorobacteriaceae). *Hydrobiologia* **31**, 317–333.

Daniels, L., Fuchs, G., Thauer, R. K. and Zeikus, J. G. (1977). Carbon monoxide oxidation by methanogenic bacteria. *J. Bacteriol.* **132**, 118–126.

Dart, P. J. (1974). The infection process. *In* "The Biology of Nitrogen Fixation" (A. Quispel, ed.), pp. 381–429. North Holland Publishing Co., Amsterdam.

Davis, J. B. and Yarborough, H. F. (1966). Anaerobic oxidation of hydrocarbons by *Desulfovibrio desulfuricans. Chem. Geol.* **1**, 137–144.

Dawson, R. N. and Murphy, K. L. (1972). The temperature dependency of biological denitrification. *Water Res.* **6**, 71–83.

Day, J. M., Neves, M. C. P. and Döbereiner, J. (1975). Nitrogenase activity on the roots of tropical forage grasses. *Soil Biol. Biochem.* **1**, 107–112.

Deelman, J. C. (1975). Bacterial sulfate reduction affecting carbonate sediments. *Soil Sci.* **119**, 73–80.

Deepe, K. and Engel, H. (1960) Untersuchungen über die temperaturabhängkeit der Nitratbildung durch *Nitrobacter winogradski* Buch. bei ungehemmten und gehemmten Wachstum. *Zentbl. Bakt. Parasitkde.* II. **113**, 561–568.

Degens, E. T. (1970) Molecular nature of nitrogenous compounds in sea water and recent marine sediments. *In* "Organic Matter in Natural Water" (D. W. Hood, ed.). Occ. Publ. 1. *Inst. Mar. Sci.,* Fairbanks, Alaska.

Degens, E. T. and Mopper, K. (1975). Early diagenesis of organic matter in marine soils. *Soil Sci.* **119**, 65–72.

Degens, E. T. and Ross, D. A. (1969). "Hot Brines and Recent Heavy Metal Deposits in the Red Sea". Springer-Verlag, New York.

Degens, E. T. and Stoffers, P. (1976). Stratified waters as a key to the past. *Nature* **263**, 22–27.

Delwiche, C. C. (1970). The nitrogen cycle. *Sci. Amer.* **223:3**, 136–146.

Delwiche, C. C. and Bryan, B. B. (1976). Denitrification. *Ann. Rev. Microbiol.* **30**, 241–262.

Denmead, O. T., Simpson, J. R. and Freney, J. R. (1974). Ammonia flux into the atmosphere from a grazed pasture. *Science* **185**, 609–610.

Deuser, W. G. (1970). Carbon-13 in Black Sea Waters and implications for the origin of hydrogen sulfide. *Science* **168**, 1575–1577.

de Vries, W., van Wijck-Kapteyn, W. M. C. and Oosterhuus, S. K. H. (1974). The presence and function of cytochromes in *Selenomonas tuminantium, Anaerovibrio lipolytica,* and *Veillonella alcalescens. J. Gen. Microbiol.* **88,** 69–78.

Dickinson, C. H. and Pugh, G. J. F., eds. (1974). "Biology of Plant Litter Decomposition. Academic Press, London.

Di Salvo, L. H. (1973). Microbial ecology. *In* "Biology and Geology of Coral Reefs". Vol. II Biology I (O. A. Jones and R. Endean, eds.), pp. 1–15. Academic Press, London.

Döbereiner, J. (1974). Nitrogen-fixing in the rhizosphere. *In* "The Biology of Nitrogen Fixation". (A.Quispel, ed.), pp. 86–120. North Holland Publishing Company, Amsterdam.

Doetsch, R. N. and Cook, T. M. (1973). "Introduction to Bacteria and their Eco-biology". Medical and Technical Publishing Company, Lancaster, England.

Dolin, M. I. (1961). Survey of microbial electron transport mechanisms. *In* "The Bacteria". Vol. II (I. C. Grunsalus and R. Y. Stanier, eds.), pp. 319–363. Academic Press, New York.

Dondero, N. C. (1975). The *Sphaerotilus-Leptothrix group. Ann. Rev. Microbiol.* **29,** 407–428.

Dougherty, R. W., ed. (1965). "Physiology of Digestion in the Ruminant". Butterworths, Washington

Dubinina, G. A. (1970). Untersuchungen über die Morphologie von *Metallogenium* und die Beziehungen zu *Mycoplasma. Ztschr. Allg. Mikrobiol* **10,** 309–320.

Edberg, N. and v. Hofsten, B. (1973). Oxygen uptake of bottom sediments studied *in situ* and in the laboratory. *Water Res.* **1,** 1285–1294.

Eglinton, G. and Calvin, M. (1967). Chemical fossils. *Sci. Amer.* **216:1,** 32–43.

Eglinton, G. and Murphy, M. T. J., eds. (1969). "Organic Geochemistry. Methods and Results". Springer-Verlag, Berlin.

Ehrlich, H. L. (1966). Reactions with manganese by bacteria from ferromanganese nodules. *Develop. Ind. Microbiol.* **1,** 43–60.

Enebo, L. (1951). On three bacteria connected with thermophilic cellulose fermentation. *Physiol Plant.* **4,** 652–666.

Engel, M. S. and Alexander, M. (1958). Growth and autotrophic metabolism of *Nitrosomonas europea. J. Bacteriol.* **76,** 217–222.

Eriksson, E. (1959). Atmospheric chemistry. *Svensk. Kem. Tidskr.* **71,** 15–32 (in Swedish).

Evans, M. C. W. (1975). The mechanism of energy conversion in photosynthesis. *Sci. Prog. Oxf.* **62,** 543–558.

Eylar, O. R. and Schmidt, E. L. (1959). A survey of heterotrophic microorganisms from soil for ability to form nitrite and nitrate. *J. Gen. Microbiol.* **22,** 473–481.

Fedorova, R. I., Milekhina, E. I. and Il'Yukhina, N. I. (1973). Evaluation of the method of "gas metabolism" for detecting extraterrestrial life. Identification of nitrogen-fixing organisms. *Izv. Akad. Nauk. SSSR, ser. biol.* **1973 (b),** 797–806.

Fell, J. W. and Master, I. M. (1973). Fungi associated with the degradation of mangrove (*Rhizophora mangle L.*) leaves in South Florida. *In* "Estuarine Microbial Ecology" (C. L. H. Stevenson and R. R. Colwell, eds.), pp. 455–465. University of South Carolina Press, Columbia, S.C.

Fenchel, T. (1969). The ecology of marine microbenthos IV. Structure and function of the benthic ecosystem, its chemical and physical factors and the microfauna communities with special reference to the ciliated protozoa. *Ophelia* **6,** 1–182.

Fenchel, T. (1970). Studies on the decomposition of organic detritus derived from the turtle grass *Thalassia testudinum. Limnol. Oceanogr.* **15,** 14–20.

Fenchel, T. (1972). Aspects of decomposer food chains in marine benthos. *Verh. deutsch. zool. Ges.* 65 *Jahresversamml.*, 14–22.

Fenchel, T. (1977). The significance of bacterivorous protozoa in the microbial community of detrital particles. *In* "Aquatic Microbial Communities.. (J. Cairns, ed.), pp. 529–544. Garland Publishing Inc., New York.

Fenchel, T. and Harrison, P. (1976). The significance of bacterial grazing and mineral cycling for the decomposition of particulate detritus. *In* "The role of Terrestrial and Aquatic Organisms in the Decomposition Processes" (J. M. Anderson and A. Macfadyen, eds.), pp. 285–299. Blackwell, Oxford.

Fenchel, T. and Jørgensen, B. B. (1977). Detritus food chains of aquatic ecosystems: the role of bacteria. *Adv. Microbial Ecol.* 1, 1–57.

Fenchel, T., Kofoed, L. H. and Lappalainen, A. (1975). Particle size-selection of two deposit feeders: the amphipod *Corophium volutator* and the prosobranch *Hydrobia ulvae*. *Mar. Biol.* 30, 119–128.

Fenchel, T., Perry, E. and Thane, A. (1977). Anaerobiosis and symbiosis with bacteria in free-living ciliates. *J. Protozool.* 24, 154–163.

Fenchel, T. and Riedl, R. J. (1970). The sulfide system: a new biotic community underneath the oxidized layer of marine sand bottoms. *Mar. Biol.* 7, 255–268.

Ferguson, J., Bubela, B. and Davies, P. J. (1975). Simulation of sedimentary ore-forming processes: concentration of Pb and Zn from brines into organic and Fe-bearing carbonate sediments. *Geologische Rundschau* 64, 767–782.

Ferry, J. G. and Wolfe, R. S. (1976). Anaerobic degradation of benzoate to methane by a microbial consortium. *Arch. Microbiol.* 107, 33–40.

Flühler, H., Ardakani, M. S., Szuskiewicz, T. E. and Stolzy, L. H. (1976). Field-measured nitrous oxide concentrations, redox potentials, oxygen diffusion rates and oxygen partial pressures in relation to denitrification. *Soil Sci.* 122, 107–114.

Floodgate, G. D. (1972). Biodegradation of hydrocarbons in the sea. *In* "Water Pollution Microbiology" (R. Mitchell, ed.), pp. 153–171. John Wiley & Sons, New York.

Focht, D. D. (1974). The effects of temperature, pH and aeration on the production of nitrous oxide and gaseous nitrogen—a zero-order kinetic model. *Soil Sci.* 118, 173–179.

Focht, D. D. and Verstraete, W. (1977). Biochemical ecology of nitrification and denitrification. *Adv. Microbial Ecol.* 1, 135–214.

Fox, C. E., Magrum, L. J., Balch, W. E., Wolfe, R. S. and Woese, C. R. (1977). Classification of methanogenic bacteria by 16S ribosomal RNA characterization. *Proc. Natl. Acad. Sci.* 74, 4537–4541.

French, J. R. J., Turner, G. L. and Bradbury, J. F. (1976). Nitrogen fixation by bacteria from the hind gut of termites. *J. Gen. Microbiol.* 95, 202–206.

Friedman, G. M. (1972). Significance of Red Sea in problem of evaporites and basinal limestones. *Am. Assoc. Petrol. Geol. Bull.* 56, 1072–1086.

Fuhs, G. W. (1961). Der mikrobielle Abbau von Kohlenwasserstoffen. *Arch. Mikrobiol.* 39, 374–422.

Fuhs, G. W., Demmerle, D. D., Canelli, E. and Chen, M. (1972). Characterization of phosphorus limited plankton algae. *In* "Nutrients and Eutrophication: The Limiting-Nutrient Controversy" (G. E. Likens, ed.), pp. 113–133. Special Symposia I. Amer. Soc. Limnol. Oceanogr. Inc., Allen Press, Lawrence, Kansas.

Garcia, J-L. (1974). Réduction de l'oxide nitreux dans les sols de rizières du Sénégal: mesure de l'activité dénitrifiante. *Soil Biol. Biochem.* 6, 79–84.

Garcia, J-L. (1975). Evaluation de la dénitrification dans les rizières par la methode de réduction de N_2O. *Soil Biol. Biochem.* 7, 251–256.

Garlick, S., Oren, A. and Padan, E. (1977). Occurrence of facultative anoxygenic photosynthesis among filamentous and unicellular cyanobacteria. *J. Bacteriol.* **129**, 623–629.

Gersberg, R., Krohn, K., Peek, N. and Goldman, C. R. (1976). Denitrification studies with ^{13}N-labelled nitrate. *Science* **192**, 1229–1231.

Gibson, D. T. (1968). Microbial degradation of aromatic compounds. *Science* **161**, 1093–1097.

Gjessing, E. T. (1976). "Physical and Chemical Characteristics of Aquatic Humus". Ann Arbor Science, Ann Arbor, Michigan.

Goering, J. J. and Dugdale, R. C. (1966a). Denitrification rates in an island bay in the Equatorial Pacific Ocean. *Science* **154**, 505–506.

Goering, J. J. and Dugdale, V. A. (1966b). Estimates of the rate of denitrification in a subarctic lake. *Limnol Oceanogr.* **11**, 113–117.

Goldhaber, M. B. and Kaplan, I. R. (1974). The sulfur cycle. *In* "The Sea" 5 (E. D. Goldberg, ed.), pp. 569–655. John Wiley & Sons, New York.

Gordon, G. C., Robinson, G. G. C., Hondzel, L. L. and Gillespie, D. C. (1973). A relationship between heterotrophic utilization of organic acids and bacterial populations in West Blue Lake, Manitoba. *Limnol Oceanogr.* **18**, 264–269.

Graetz, D. A., Keeney, D. R. and Aspiras, R. B. (1973). Eh status of lake sediment-water systems in relation to nitrogen transformations. *Limnol. Oceanogr.* **18**, 908–917.

Gray, T. R. G., Baxby, P., Hill, I. R. and Goodfellow, M. (1968). Direct observation of bacteria in soil *In* "The Ecology of Soil Bacteria" (T. R. G. Gray and D. Parkinson, eds.), pp. 171–197. University of Toronto Press, Toronto.

Gray, T. R. G. and Parkinson, D., eds. (1968). "The Ecology of Soil Bacteria". University of Toronto Press, Toronto.

Gray, T. R. G. and Williams, S. T. (1971). "Soil Microorganisms". Oliver and Boyd, Edinburgh.

Gundersen, K. and Mountain, C. W. (1973). Oxygen utilization and pH change in the ocean resulting from biological nitrate formation. *Deep Sea Res.* **20**, 1083–1091.

Guyer, M. and Hegeman, G. (1969). Evidence for a reductive pathway for the anaerobic metabolism of benzoate. *J. Bact.* **99**, 906–907.

Hadjipetrou, L. P. and Stouthamer, A. H. (1965). Energy production during nitrate respiration by *Aerobacter aerogines. J. Gen. Microbiol.* **38**, 29–34.

Hairston, N. G., Smith, F. E. and Slobodkin, L. B. (1960). Community structure, population control and competition. *Amer. Nat.* **94**, 421–425.

Hall, D. O., Commack, R. and Rao, K. K. (1973). The plant ferredoxins and their relationship to the evolution of ferredoxins from primitive life. *Pure Appl. Chem.* **34**, 553–575.

Hall, J. B. (1971). Evolution of the prokaryotes. *J. Theoret. Biol.* **30**, 429–454.

Hall, J. B. (1973). The nature of the host in the origin of the eukaryotic cell. *J. Theoret. Biol.* **38**, 413–418.

Hall, K. J., Kleiber, P. M. and Ysaki, I. (1972). Heterotrophic uptake of organic solutes by microorganisms in the sediment. *Mem. Ist. Ital. Idrobiol.* **44**, 441–471.

Hallberg, R. O. (1972). Sedimentary sulfide formation—an energy circuit approach. *Mineral Deposita (Berl.)*, **7**, 189 201.

Hamilton, R. D. and Preslan, J. E. (1970). Observations on heterotrophic activity in the eastern tropical Pacific. *Limnol. Oceanogr.* **15**, 395–401.

Hansen, M. H., Ingvorsen, K. and Jørgensen, B. B. (1978). Mechanisms of hydrogen sulfide release from coastal marine sediments to the atmosphere. *Limnol. Oceanogr.* **23**, 68–76.

Hansen, T. A. and Van Gemerden, H. (1972). Sulfide utilization by purple sulfur bacteria. *Arch. Microbiol.* **86**, 49–56.

Hargrave, B. T. (1970). The effect of a deposit-feeding amphipod on the metabolism of benthic microflora. *Limnol. Oceanogr.* **15**, 21–30.

Hargrave, B. T. (1971). An energy budget for a deposit-feeding amphipod. *Limnol. Oceanogr.* **16**, 99–103.

Hargrave, B. T. (1976). The central role of invertebrate faeces in sediment decompossition. *In* "The Role of Terrestrial and Aquatic Organisms in Decomposition Processes" (J. M. Anderson and A. Macfadyen, eds.), pp. 301–321. Blackwell, Oxford.

Hasan, S. M. and Hall, J. B. (1975). The physiological function of nitrate reduction in *Clostridium perfringens. J. Gen. Microbiol.* **87**, 120–128.

Hathaway, J. C. and Degens, E. T. (1969). Methane-derived marine carbonates of Pleistocene age. *Science* **165**, 690–692.

Hauck, R. D. and Bremner, J. M. (1976). Use of tracers for soil and fertilizer research. *Adv. Agron.* **28**, 219–266.

Head, W. D. and Carpenter, E. J. (1975). Nitrogen fixation associated with the marine macro-alga *Codium fragile. Limnol. Oceanogr.* **20**, 815–823.

Heyer, J. and Schwartz, W. (1970). Untersuchungen zur Erdölmikrobiologie v. Leben in nicht-wässrigen Medien. 1. Verhalten von Mikroorganismen in Kontakt mit Mineralöl. *Ztschr. Allg. Mikrobiol.* **10**, 545–563.

Hirsch, P. and Engel, H. (1965). Über oligocarbophile Actinomyceten. *Ber. Dtsch. Bot. Ges.* **69**, 441–454.

Hobbie, J. E. (1967). Glucose and acetate in freshwater: concentrations and turnover rates. *In* "Chemical Environment in Aquatic Habitat" (H. L. Golterman and R. S. Clymo, eds.), pp. 245–251. North Holland Publishing Company, Amsterdam.

Hobbie, J. E. (1971). Heterotrophic bacteria in aquatic ecosystems; some results of studies with organic radio isotopes. *In* "The Structure and Function of Fresh-Water Microbial Communities" (J. Cairns, Jr., ed.), pp. 181–194. Research Div Monograph 3, Virginia Polytechnic Institute and State University, Blacksbury, Va.

Hobbie, J. E. and Crawford, C. C. (1969). Respiration corrections for bacterial uptake of dissolved organic compounds in natural water. *Limnol. Oceanogr.* **14**, 528–532.

Hobbie, J. E., Holm-Hansen, O., Packard, T. T., Pomeroy, L. R., Sheldon, R. W., Thomas, J. P. and Wiebe, W. J. (1972). A study of the distribution and activity of microorganisms in ocean water. *Limnol. Oceanogr.* **17**, 544–555.

Hofsten, B.v. and Edberg, N. (1972). Estimating the rate of degradation of cellulose filters in water. *Oikos.* **23**, 29–34.

Holm-Hansen, O. and Booth, C. R. (1966). The measurement of adenosine triphosphate in the ocean and its ecological significance. *Limnol. Oceanogr.* **11**, 510–519.

Hopwood, A. P. and Downing, A. L. (1965). Factors affecting the rate of production and properties of activated sludge in plants treating domestic sewage. *J. Proc. Inst. Sew. Purif.* pp. 435–448.

Horowitz, N. H. (1965). The evolution of biochemical synthesis—retrospect and prospect. *In* "Evolving Genes and Proteins" (V. Bryson and H. J. Vogel, eds.), pp. 15–23. Academic Press, New York.

Hullah, W. A. and Blackburn, T. H. (1971). Uptake and incorporation of amino acids and peptides by *Bacteroides amylophilus. Appl. Microbiol.* **21**, 187–191.

Hungate, R. E. (1955). Mutualistic intestinal protozoa. *In* "Biochemistry and Physiology of Protozoa" (S. H. Hutner and A. Lwoff, eds.), pp. 159–199. Academic Press, New York.

Hungate, R. E. (1963). Symbiotic associations: the rumen bacteria. *Symposia Soc. Gen. Microbiol.* **13**, 266–297.

Hungate, R. E. (1966). "The Rumen and its Microbes". Academic Press, New York.

Hungate, R. E. (1975). The rumen microbial ecosystem. *Ann. Rev. Ecol. Syst.* **6**, 39–66.

Hutchinson, G. E. (1954). The biochemistry of the terrestrial atmosphere. *In* "The Solar System". Vol. 2 (G. Kuiper, ed.), pp. 371–433. The University of Chicago Press, Chicago.

Hynes, H. B. N. and Kaushik, N. K. (1969). The relationships between dissolved nutrient salts and protein production in submerged autumnal leaves. *Verh. Internat. Verein. Limnol.* **17**, 95–103.

Ianotti, E. L., Kafkewitz, D., Wolin, M. J. and Bryant, M. P. (1973). Glucose fermentation products of *Ruminococcus albus* grown in continuous culture with *Vibrio succinogenes:* changes caused by interspecies transfer of hydrogen. *J. Bacteriol.* **114**, 1231–1240.

Ivanov, M. V. (1968). "Microbiological Processes in the Formation of Sulfur Deposits". Israel Program for Scientific Translations, Jerusalem.

Jackson, T. A. and Moore, C. B. (1976). Secular variations in kerogen structure and carbon, nitrogen and phosphorus concentrations in pre phanerozoic and phanerozoic sedimentary rocks. *Chem. Geol.* **18**, 107–136.

Jacobsen, J. (1976). Mobilixation, transportation and sedimentation of weathering products from the abandoned brown-coal pits. (Iron pollution of the river Skjernå and Ringkøbing Fjord, Western Jutland). *Danm. Geol. Unders. Årbog* **1975**, 57–74.

Janis, C. (1976). The evolutionary strategy of Equidae and the origins of rumen and cecal digestion. *Evolution* **30**, 757–774.

Jannasch, H. W. (1974). Steady state and the chemostat in ecology. *Limnol. Oceanogr.* **19**, 716–720.

Jannasch, H. W. and Pritchard, P. H. (1972). The role of inert particulate matter in the activity of aquatic microorganisms. *Mem. Ist. Ital. Idrobiol.* **29**, suppl. 289–308.

Jensen, V. (1975). Bacterial flora of soil after application of oily waste. *Oikos* **26**, 152–158.

Jernelöv, A. and Martin, A.-L. (1975). Ecological implications of metal metabolism by microorganisms. *Ann. Rev. Microbiol.* **29**, 62–77.

Johannes, R. E. and Satomi, M. (1967). Measuring organic matter retained by aquatic invertebrates. *J. Fish Res. Board Can.* **24**, 2467–2471.

Johnson, A. H. and Stokes, J. L. (1966). Manganese oxidation by *Sphaerotilus discophorous. J. Bacteriol.* **91**, 1543–1547.

Jones, O. T. G. (1977). Electron transport and ATP synthesis in photosynthetic bacteria. *Symposia Soc. Gen. Microbiol.* **27**, 151–183.

Jørgensen, B. B. (1977a). The sulfur cycle of a coastal marine sediment (Limfjorden, Denmark). *Limnol. Oceanogr.* **22**, 814–832.

Jørgensen, B. B. (1977b). Distribution of colorless sulfurbacteria (*Beggiatoa* spp.) in a coastal marine sediment. *Mar. Biol.* **41**, 19–28.

Jørgensen, B. B. (1977c). Bacterial sulfate reduction within reduced microniches of oxidized marine sediments. *Mar. Biol.* **41**, 7–17.

Jørgensen, B. B. and Cohen, Y. (1977). Solar Lake (Sinai). 5. The sulfur cycle of the benthic cyanobacterial mats. *Limnol. Oceanogr.* **22**, 657–666.

Jørgensen, B. B. and Fenchel, T. (1974). The sulfur cycle of a marine sediment model system. *Mar. Biol.* **24**, 189–201.

Jørgensen, C. B. (1966). "Biology of Suspension Feeding". Pergamon Press, Oxford.

Jørgensen, C. B. (1976). Augüst Pütter, August Krogh, and modern ideas on the use of dissolved organic matter in aquatic environments. *Biol. Rev.* **51**, 291–328.

Kamp-Nielsen, L. and Andersen, J. M. (1977). A review of the literature on sediment–water exchange of nitrogen compounds. *Prog. Wat. Tech.* **8**, 393–418.

Kallio, R. E., Finnerty, W. R., Wawzonek, S. and Klimstra, P. D. (1963). Mechanisms in the microbial oxidations of alkanes. *In* "Marine Microbiology" (C. Oppenheimer, ed.), pp. 453–463. Charles C. Thomas, Springfield, Ill.

Kaplan, I. R., ed. (1974). "Natural Gases in Marine Sediments". Plenum, New York.

Kemp, A. L. W. and Thode, H. G. (1968). The mechanism of the bacterial reduction of sulfate and sulfite from isotope fractional studies. *Geochim. Cosmochim. Acta.* **32**, 71–91.

Kenyon, C. N., Rippka, R. and Stanier, R. Y. (1972). Fatty acid composition and physiological properties of some filamentous blue-green algae. *Arch. Microbiol.* **83**, 216–236.

Kerr, P. C., Brockway, D. L., Paris, D. R. and Sanders, W. M. III. (1972). The carbon cycle in aquatic ecosystems. *In* "Nutrients in Natural Waters" (H. E. Allen and J. R. Kramer, eds.), pp. 101–124. John Wiley & Sons, New York.

Knowles, G., Downing, A. L. and Barrett, M. J. (1965). Determination of kinetic constants for nitrifying bacteria in mixed cultures, with the aid of an electronic computer. *J. Gen. Microbiol.* **38**, 263–278.

Koenings, J. P. (1976). In situ experiments on the dissolved and colloidal state of iron in an acid bog lake. *Limnol. Oceanogr.* **21**, 674–683.

Koike, I. and Hattori, A. (1975). Energy yield of denitrification: an estimate from growth yield in continuous cultures of *Pseudomonas denitrificans* under nitrate-, nitrite- and nitrous oxide-limited conditions. *J. Gen. Microbiol.* **88**, 11–19.

Koike, I. and Hattori, A. (1978). Denitrification and ammonia formation in anaerobic coastal sediments. *Appl. Environ. Microbiol.* **35**, 278–282.

Koike, I., Wada, E., Tsuji, T. and Hattori, A. (1972). Studies on denitrification in a brackish lake. *Arch. Hydrobiol.* **69**, 508–520.

Koyama, T. (1963). Gaseous metabolism in lake sediments and paddy soils and the production of atmospheric methane and hydrogen. *J. Geophys. Res.* **68**, 3971–3973.

Krumbein, W. E. (1971). Manganese-oxidizing fungi and bacteria. *Die Naturwissenschaften* **58**, 56–57.

Kugelman, I. J. and Chin, K. K. (1971). Toxicity, synergism and antagonism in anaerobic waste treatment processes. *Adv. Chem. Ser.* **105**, 55–90.

Kuznetzov, S. I. (1968). Recent studies on the role of microorganisms in the cycling of substances in lakes. *Limnol. Oceanogr.* **13**, 211–224.

Kuenen, J. G. (1975). Colourless sulfur bacteria and their role in the sulfur cycle. *Plant Soil.* **43**, 49–76.

Kvenvolden, K. A. (1976). Natural evidence for chemical and early biological evolution. *Origins of Life* **5**, 71–86.

Larsen, H. (1960). Chemosynthesis (general). *In* "Handbuch der Pflanzenphysiologie". Vol. 5, 2, 613–648, Berlin.

Lasker, R. and Giese, A. C. (1954). Nutrition of the sea urchin *Stronglylocentrotus purpuratus*. *Biol. Bull.* **106**, 328–340.

Lawrence, A. W. (1971). Application of process kinetics to design of anaerobic processes. *Adv. Chem. Ser.* **105**, 163–189.

Lees, H. (1952). The biochemistry of nitrifying organisms. 1. The ammonia oxidizing systems of *Nitrosomonas*. *Biochem. J.* **52**, 134–139.

Lees, H. and Quastel, J. H. (1946). Biochemistry of nitrification in soil. II. The site of soil nitrification. *Biochem. J.* **40**, 815–823.

Lehninger, A. L. (1971). "Bioenergetics". W. A. Benjamin, Inc., New York.

Lehtomäki, M. and Niemelä, S. (1975). Improving microbial degradation of oil in soil. *Ambio* **4**, 126–129.

Leiner, M., Bhowmik, D. K., König, K. and Fisher, M. (1968). Die Gärung und Atmung von *Pelomyxa palustris* Greeff. *Biol. Zbl.* **87**, 567–591.

Lenhard, G. (1968). A standardized procedure for the determination of dehydrogenase activity in samples from anaerobic treatment systems. *Water Res.* **2**, 161–167.

Lersten, N. R. and Horner, H. T. Jr. (1976). Bacterial leaf nodule symbiosis in angiosperms with emphasis on Rubiacea and Myrsinaecae. *Bot. Rev.* **42**, 145–214.

Levy, J., Campbell, J. J. R. and Blackburn, T. H. (1973). "Introductory Microbiology". John Wiley & Sons, New York.

Lex, M., Silvester, W. and Stewart, W. D. D. (1972). Photorespiration and nitrogenase in the blue-green alga, *Anabaena cylindrica*. *Proc. Roy. Soc. B.* **180**, 87–102.

Lundgren, D. G., Vestal, J. R. and Tabita, F. R. (1972). The microbiology of mine drainage pollution. *In* "Water Pollution Microbiology" (R. Mitchell, ed.), pp. 69–88. John Wiley & Sons, New York.

Maciag, W. J. and Lundgren, D. G. (1964). Carbon dioxide fixation in the chemoautotroph *Ferrobacillus ferrooxidans*. *Biochem. Biophys. Res. Commun.* **17**, 603–607.

McComb, J. A., Elliot, J. and Dilworth, M. J. (1975). Acetylene reduction by *Rhizobium* in pure culture. *Nature* **156**, 409–410.

Mackintosh, M. E. (1971). Nitrogen fixation by *Thiobacillus ferrooxidans* species. *J. Gen. Microbiol.* **66**, i.

McLaren, A. D. and Peterson, G. H., eds (1967). "Soil Biochemistry". Marcel Dekker, Inc., New York.

McLaughlin, P. J. and Dayhoff, M. O. (1970). Eukaryotes versus prokaryotes: an estimate of evolutionary distance. *Science* **168**, 1469–1471.

Mah, R. A., Hungate, R. E. and Ohwaki, K. (1976). Acetate, a key intermediate in methanogenesis. *In* "Microbial Energy Conversion" (H. G. Schlegel and J. Barnea, eds.), pp. 97–106. Erich Glotze KG, Göttingen.

Mann, K. H. (1972). Macrophyte production and detritus food chains in coastal waters. *Mem. Ist. Ital. Idrobiol.* **29**, *Suppl.*, 353–383.

Mann, K. H. (1976). Decomposition of marine macrophytes. *In* "The role of Terrestrial and Aquatic Organisms in Decomposition Processes". (J. M. Anderson and A. Macfadyen, eds.), pp. 247–267. Blackwell, Oxford.

Mann, L. D., Focht, D. D., Joseph, H. A. and Stolzy, L. H. (1972). Increased denitrification in soils by additions of sulfur as an energy source. *J. Environ. Quality* **1**, 329–332.

Margulis, L. (1970). "Origin of Eukaryotic Cells". Yale University Press, New Haven.

Margulis, L. (1971). Symbiosis and evolution. *Sci. Amer.* **224:2**, 48–57.

Margulis, L. (1972). Early cellular evolution. *In* "Exobiology" (C. Ponnamperuma, ed.), pp. 342–368. North Holland Publishing Company, Amsterdam.

Margulis, L. (1975). The microbes' contribution to evolution. *Bio Systems* **7**, 266–292.

Margoulis, P. J. and Bandy, A. R. (1977). Estimate of the contribution of biologically produced dimethyl sulfide to the global sulfur cycle. *Science* **196**, 647–648.

Martens, C. S. (1976). Control of methane sediment-water bubble transport by macroinfaunal irregation in Cape Lookout Bight, North Carolina. *Science* **192**, 998–1000.

Martens, C. S. and Berner, R. A. (1974). Methane production in the interstitial waters of sulfate-depleted marine sediments. *Science* **185**, 1167–1169.

Martin, J. P., Haider, K., Farmer, W. J. and Fustec-Mathon, E. (1974). Decomposition and distribution of residual activity of some ^{14}C-microbial poly-saccharides and cells, glucose, cellulose and wheat straw in soil. *Soil. Biol. Biochem.* **6**, 221–230.

Mechalas, B. J. (1974). Pathways and environmental requirements for biogenic gas production in the oceans. *In* "Natural Gases in Marine Sediments" (I. R. Kaplan, ed.), pp. 12–25. Plenum Press, New York.

Medveczky, N. and Rosenberg, H. (1971). Phosphate transport in *Escherichia coli*. *Biochem. Biophys. Acta* **241**, 404–506.

Meiklejohn, J. (1954). Some aspects of the physiology of the nitrifying bacteria. *Symposia Soc. Gen. Microbiol.* **4**, 68–83.

Meyers, S. P. and Hopper, B. E. (1973). Nematological–microbial interrelationships and estuarine biodegradative processes. *In* "Estuarine Microbial Ecology" (L. H. Stevenson and R. R. Colwell, eds.), pp. 483–489. University of South Carolina Press, Columbia S. C.

Millbank, J. W. (1974). Associations with blue-green algae. *In* "The Biology of Nitrogen Fixation" (A. Quispel, ed.), pp. 238–264. North Holland Publishing Company, Amsterdam.

Miller, S. L. and Orgel, L. E. (1974). "The Origins of Life on Earth". Prentice-Hall, In., Englewood Cliffs, New Jersey.

Mitchell, P. (1977). Epilogue: from energetic abstraction to biochemical mechanism. *Symposia Soc. Gen. Microbiol.* **27**, 383–423.

Moir, R. J. (1965). The comparative physiology of ruminant-like animals. *In* "Physiology of Digestion in the Ruminant" (R. W. Dougherty, ed.), pp. 1–23. Butterworths, Washington.

Monheimer, R. H. (1974). Sulfate uptake as a measure of planktonic microbial production in freshwater ecosystems. *Can. J. Microbiol.* **20**, 825–831.

Moore, S. F. and Schroeder, E. D. (1971). The effect of nitrate feed rate on denitrification. *Water Res.* **5**, 445–452.

Morita, R. S. (1968). Evidence for three photochemical systems in *Chromatium* D. *Biochem. Biophys. Acta.* **153**, 241.

Mortimer, C. H. (1941–42). The exchange of dissolved substances between mud and water in lakes. I-II, *J. Ecol.* **29**, 280–329; **30**, 147–201.

Mosser, J. L., Mosser, A. G. and Brock, T. D. (1973). Bacterial origin of sulfuric acid in geothermal habitats. *Science* **179**, 1323–1324.

Mulder, E. G. (1964). Iron bacteria, particularly those of the *Sphaerotilus-Leptothrix* group, and industrial problems. *J. Appl. Bact.* **27**, 151–173.

Munro, A. L. S. and Brock, T. D. (1968). Distinction between bacterial and algal utilization of soluble substances in the sea. *J. Gen. Microbiol.* **51**, 35–41.

Nedwell, D. B. (1975). Inorganic nitrogen metabolism in a eutrophicated tropical mangrove estuary. *Water Res.* **9**, 221–231.

Nedwell, D. B. and Floodgate, G. D. (1972). Temperature-induced changes in the formation of sulfide in a marine sediment. *Mar. Biol.* **14**, 18–24.

Nömmik, H. (1976). Predicting the nitrogen-supplying power of acid forest soils from data on the release of CO_2 and NH_3 on partial oxidation. *Commun. Soil Sci. Analysis* **7**, 569–584.

Nottingham, P. M. and Hungate, R. E. (1969). Methanogenic fermentation of benzoate. *J. Bact.* **98**, 1170–1172.

Nutman, P. S. (1963). Factors influencing the balance of mutual advantage in legume symbiosis. *Symposia Soc. Gen. Microbiol.* **13**, 51–71.

Nutman, P. S., ed. (1976). "Symbiotic Nitrogen Fixation in Plants". Cambridge University Press, Cambridge.

Odum, E. P. and de la Cruz, A. A. (1967). Particulate organic detritus in a Georgia salt marsh-estuarine ecosystem. *In* "Estuaries" (G. H. Lauff, ed.), pp. 383–388. Publ. Amer. Assoc. Adv. Sci., **83**.

Oehler, D. Z. and Schopf, J. W. (1972). Carbon isotopic studies of organic matter in Precambrian rocks. *Science* **175**, 1246–1248.

Ogata, G. and Bower, C. A. (1965). Significance of biological sulfate reduction on soil acidity. *Soil Sci. Soc. Amer. Proc.* **29**, 23–25.

Ogura, N. (1975). Further studies on decomposition of dissolved organic matter in coastal seawater. *Mar. Biol.* **31**, 101–111.

Ohle, W. (1962). The metabolism of lakes as a basis for a general metabolic dynamism. *Kiel Meeresforsch* **18**, 107–120.

Osterhelt, D., Gottschlik, R., Hartman, R., Michel, H. and Wagner, G. (1977). Light energy conversion in halobacteria. *Symposia Soc. Gen. Microbiol.* **27**, 333–349.

Olah, J. (1972). Leaching, colonization and stabilization during detritus formation. *Mem. Ist. Ital. Idrobiol.* **29**, *Suppl.*, 105–127.

Olsen, C. (1970). On biological nitrogen fixation in nature, particularly in blue-green algae. *C.R. trav. lab. Carlsberg* **37**, 269–283.

Olson, J. M. (1970). The evolution of photosynthesis. *Science* **168**, 438–446.

Oremland, R. S. (1976). Methane production in shallow-water tropical marine sediments. *Appl. Microbiol.* **30**, 602–608.

Overbeck, J. (1974). Microbiology and biochemistry. *Mitt. Internal Verein. Limnol.* **20**, 198–228.

Owen, T. and Biemann, K. (1976). Composition of the atmosphere at the surface of Mars: detection of argon 36 and preliminary analysis. *Science* **193**, 801–802.

Otsuki, A. and Hanya, T. (1972). Production of dissolved organic matter from dead green algal cells. I. Aerobic microbial decomposition. *Limnol. Oceanogr.* **17**, 248–257.

Paerl, H. W. (1973). Detritus in Lake Tahoe, structural modification by attached microflora. *Science* **180**, 496–498.

Paerl, H. W. (1974). Bacterial uptake of dissolved organic matter in relation to detrital aggregation in marine and freshwater systems. *Limnol. Oceanogr.* **19**, 966–972.

Paerl, H. W. and Goldman, C. R. (1972). Stimulation of heterotrophic and autotrophic activities of a planktonic microbial community by siltation at Lake Tahoe, California. *Mem. Ist. Ital. Idrobiol.* **29**, *Suppl.* 129–147.

Painter, H. W. (1970). A review of literature on inorganic nitrogen metabolism in microorganisms. *Water Res.* **4**, 393–450.

Pamatmat, M. M. (1971). Oxygen consumption by the seabed. 4. Shipboard and laboratory experiments. *Limnol. Oceanogr.* **16**, 536–550.

Park, R. and Epstein, S. (1960). Carbon isotopic fractionation during photosynthesis. *Geochim. Cosmochim. Acta.* **21**, 110–126.

Parson, T. R. (1963). Suspended organic matter in sea water. *In* "Progress in Oceanography". Vol. 1 (M. Sears, ed.), pp. 205–239. Pergamon Press, Oxford.

Patrick, W. H. Jr. and Reddy, K. R. (1976). Nitrification–denitrification reactions in flooded soils and water bottoms: dependence on oxygen supply and ammonia diffusion. *J. Environ. Qual.* **5**, 469–472.

Payne, W. J. (1973). Reduction of nitrogenous oxides by microorganisms. *Bact. Rev.* **37**, 409–452.

Pearl, I. A. (1967). "The Chemistry of Lignin". Marcel Dekker, Inc., New York.

Peck, H. D. (1974). The evolutionary significance of inorganic sulfur metabolism. *Symposia Soc. Gen. Microbiol.* **24**, 241–262.

Petersen, C. G. J. (1918). The sea bottom and its production of fish food. A survey of work done in connection with the valuation of the Danish waters from 1883–1917. *Rep. Danish Biol. Sta.* **21**, 1–64; *Append.* 1–68.

Pfennig, N. (1967). Photosynthetic bacteria. *Ann. Rev. Microbiol.* **21**, 285–324.

Pfennig, N. (1975). The phototrophic bacteria and their role in the sulfur cycle. *Plant Soil* **43**, 1–16.

Pfennig, N. and Biebl, H. (1976). *Desulfuromonas acetoxidans* gen. nov and sp. nov., a new anaerobic sulfur-reducing, acetate-oxidizing bacterium. *Arch. Microbiol.* **110**, 3–12.

Pierson, B. K. and Castenholz, R. W. (1971). Bacterio-chlorophylls in gliding filamentous prokaryotes from hot springs. *Nature* **233**. 25–27.

Pine, M. J. (1971). The methane fermentations. *Adv. Chem. Ser.* **105**, 1–10.

Pittman, K. A., Lakshmann, S. and Bryant, M. (1967). Oligopeptide uptake by *Bacteroides tuminicola*. *J. Bacteriol.* **93**, 1499–1508.

Postgate, J. R. (1969). The sulfur cycle. *In* "Inorganic Sulfur Chemistry" (G. Nickless, ed.), pp. 259–279. Elsevier, Amsterdam.

Postgate, J. R. (1974). Evolution within nitrogen-fixing systems. *Symposia Soc. Gen. Microbiol.* **24**, 263–292.

Potrikus, C. J. and Breznak, J. A. (1977). Nitrogen-fixing *Enterobacter agglomerans* isolated from the guts of wood-eating termites. *Appl. Environ. Microbiol.* **33**, 392–399.

Pratt, P. F., Davis, S. and Sharpless, R. G. (1976). A four-year trial with animal manures. *Hilgardia*. **44**, 99–125.

Prim, P. and Lawrence, J. M. (1975). Utilization of marine plants and their constituents by bacteria isolated from the gut of echinoids (Echinodermata). *Mar. Biol.* **33**, 167–173.

Pringsheim, E. G. (1949). Iron bacteria. *Biol. Rev.* **24**, 200–245.

Quale, J. R. (1972). The metabolism of one-carbon compounds by microorganisms. *Adv. Microbial. Physiol.* **7**, 119–203

Quispel, A., ed. (1974). "The Biology of Nitrogen Fixation". North Holland Publishing Company, Amsterdam.

Raff, R. A. and Mahler, H. R. (1972). The non symbiotic origin of mitochondria. *Science* **177**, 575–582.

Ramm, A. E. and Bella, D. A. (1974). Sulfide production in anaerobic microcosms. *Limnol. Oceanogr.* **19**, 110–118.

Reddy, K. R. and Patrick, W. H. (1976). Effect of frequent changes in aerobic and anaerobic conditions on redox potential and nitrogen loss in a flooded soil. *Soil Biol Biochem.* **8**, 491–495.

Reeburgh, W. S. and Heggie, D. T. (1975). Depth distribution of gases in shallow water sediments. *In* "Natural Gases in Marine Sediments" (I. R. Kaplan, ed.), pp. 27–45. Plenum, New York.

Rees, C. E. (1970). The sulfur isotope balance of the ocean: an improved model. *Earth Planetary Sci. Lett.* **7**, 366–370.

Rhee, A. (1972). Competition between an algae and an aquatic bacterium for phosphate. *Limnol. Oceanogr.* **17**, 505–514.

Rheinheimer, G. (1974). "Aquatic Microbiology". John Wiley & Sons, London.

Ribbons, D. W., Harrison, J. E. and Wadinski, A. M. (1970). Metabolism of single carbon compounds. *Ann. Rev. Microbiol.* **24**, 135–158.

Richards, B. N. (1976). "Introduction to the Soil Ecosystem". Longman, London.

Richards, F. A. and Broenkow, W. W. (1971). Chemical changes, including nitrate reduction in Darwin Bay, Galapagos Archipelago, over a 2 month period, 1969. *Limnol. Oceanogr.* **16**, 758–765.

Riley, G. A. (1970). Particulate organic matter in sea water. *Adv. Mar. Biol.* **8**, 1–118.

Rogers, S. R. and Anderson, J. J. (1976a). Measurement of growth and iron deposition in *Sphaerotilus discophorus*. *J. Bacteriol.* **126**, 257–263.

Rogers, S. R. and Anderson, J. J. (1976b). Role of iron deposition in *Sphaerotilus discophorus*. *J. Bacteriol.* **126**, 264–271.

Romer, A. S. (1958). "Vertebrate Paleontology". The University of Chicago Press, Chicago, Ill.

Rose, A. H. (1968). "Chemical Microbiology". Butterworths, London.

Rosswall, T. (1976). The internal nitrogen cycle between vegetation, microorganisms and soil. *In* "Nitrogen, Phosphorus and Sulfur—Global Cycles (B. H. Svensson and R. Söderlund, eds.), pp. 157–167. *Ecol. Bull. (Stockholm)* **22**.

Rozanov, A. G., Volkov, I. I., Zhabina, N. N. and Yagodinsky, T. A. (1971). Hydrogen sulfide in the sediments of the continental slope, Northwest Pacific Ocean. *Geochemistry International,* 333–339.

Rudd, J. W. M., Furutami, A., Flett, R. J. and Hamilton, R. D. (1976). Factors controlling methane oxidation in shield lakes; the role of nitrogen fixation and oxygen concentration. *Limnol. Oceanogr.* **21**, 357–364.

Rudd, J. W. M. and Hamilton, R. D. (1975). Factors controlling rates of methane oxidation and distribution of methane oxidizers in a small stratified lake. *Arch. Hydrobiol.* **75**, 522–538.

Rudd, J. W. M., Hamilton, R. D. and Campbell, N. E. R. (1974). Measurement of microbial oxidation in lake water. *Limnol. Oceanogr.* **19**, 519–524.

Russell-Hunter, W. D. (1970). "Aquatic Productivity: an Introduction to some Basic Aspects of Biological Oceanography and Limnology". Macmillan, London.

Rutten, M. G. (1970). The history of atmospheric oxygen. *Space Life Sci.* **1**, 1–13.

Satter, L. D., Suttie, J. W. and Baumgardt, B. R. (1964). Dietary induced changes in volatile fatty acid formation from α-cellulose-^{14}C and hemicellulose-^{14}C. *J. Dairy Sci.* **47**, 1365–1370.

Saunders, G. W. (1976). Decomposition in freshwater. *In* "The role of Terrestrial and Aquatic Organisms in Decomposition Processes". (J. M. Anderson and A. Macfadyen, eds.), pp. 341–373. Blackwell Scientific Publications, Oxford.

Scheitfinger, C. C., Linehab, B. and Wolin, M. J. (1975). H_2 production by *Selenomonas ruminantium* in the absence and presence of methanogenic bacteria. *Appl. Microbiol.* **29**, 480–483.

Schlegel, H. G. (1975). Mechanisms of chemoautotrophy. *In* "Marine Ecology" (O. Kinne, ed.), **2**, 9–60. John Wiley, London.

Schnepf, E. and Brown, R. M. Jr. (1971). On relationships and endosymbiosis and the origin of plastids and mitochondria. *In* "Results and Problems in Cell Differentiation", **2**, 299–322. Springer-Verlag, New York–Heidelberg–Berlin.

Schopf, J. W. (1970). Precambrian microorganisms and evoluationary events prior to the origin of vascular plants. *Biol. Rev.* **45**, 319–352.

Schopf, J. W. (1974). Palaebiology of the Precambrian: the age of the blue-green algae. *In* "Evolutionary Biology" (T. Dobzansky and M. K. Hecht, eds.), **7**, 1–43. Plenum Press, New York and London.

Schopf, J. W. (1976). Are the oldest fossils fossils? *Origins of Life* **7**, 19–36.

Schweifurth, R. (1973). Manganoxydierende Bakterien. I. Isolierung und Bestimmung einiger Stämme von Manganbakterien. *Ztschr. Allg. Mikrobiol.* **13**, 341–347.

Seki, H. (1972). The role of microorganisms in the marine food chain with reference to organic aggregation. *Mem. Ist. Ital. Idrobiol.* **29**, *Suppl.* 245–259.

Senez, J. and Azoulay, E. (1963). Pathway of lower alkane oxidation by pseudomonads. *In* "Marine Microbiology" (C. Oppenheimer, ed.), pp. 464–474. Charles C. Thomas, Springfield, Illinois.

Shapiro, J. H. (1973). Blue-green algae: why they become dominant. *Science* **179**, 382–384.

Sharp, J. H. (1969). Blue-green algae and carbonates—*Schizothrix calciola* and algal stromatolites from Bermuda. *Limnol. Oceanogr.* **14**, 568–578.

Sillén, L. G. (1966). Regulation of O_2, N_2 and CO_2 in the atmosphere: thoughts of a laboratory chemist. *Tellus* **18**, 198–206.

Silver, W. S. and Postgate, J. R. (1973). Evolution of asymbiotic nitrogen fixation. *J. Theor. Biol.* **40**, 1–10.

Silverman, M. P. and Ehrlich, H. L. (1964). Microbial formation and degradation of minerals. *Adv. Appl. Microbiol.* **6**, 153–206.

Skerman, V. B. D. (1967). "The Genera of Bacteria". 2nd edition. The Williams and Wilkins Co., Baltimore.

Skinner, F. A. and Walker, N. (1961). Growth of *Nitrosomonas europea* in batch and continuous culture. *Arch. Mikrobiol.* **38**, 339–349.

Smith, A. J., London, J. and Stanier, R. Y. (1967). Biochemical basis of obligate autotrophy in blue-green algae and thiobacilli. *J. Bacteriol.* **94**, 972–983.

Smith, J. W., Schopf, J. W. and Kaplan, I. R. (1970). Extractable organic matter in Precambrian cherts. *Geochim. Cosmochim. Acta.* **9**, 659–675.

Smith, M. P. W., Evans, M. R. and Baines, S. (1976). Simultaneous nitrification and denitrification in completely mixed reactors during treatment of animal wastes. *Proc. Soc. Gen. Microbiol.* **3**, 118.

Smith, P. H. and Mah, R. A. (1966). Kinetics of acetate metabolism during sludge digestion. *Appl. Microbiol.* **14**, 368–371.

Söderlund, R. and Svensson, R. H. (1976). The global nitrogen cycle. *In* "Nitrogen, Phosphorus and Sulfur—Global Cycles" (B. H. Svensson and R. Söderlund, eds.). *Ecol. Bull.* **22**, 23–73. Swedish Natural Science Council, Stockholm.

Sokolova, G. A. and Karavaiko, G. I. (1964). "Physiology and Geochemical Activity of Thiobacilli". Israel Program for Scientific Translations, Jerusalem 1968.

Sørensen, J. (1978a). Capacity for denitrification and reduction of nitrate to ammonia in a coastal marine sediment. *Appl. Environ. Microbiol.* **35**, 301–305.

Sørensen, J. (1978b). Denitrification rates in a marine sediment by the acetylene in hibition technique. *Appl. Environ. Microbiol.* **36**, 139–143.

Sørensen, L. H. (1962). Decomposition of lignin by soil bacteria and complex formation between auto-oxidized lignin and organic nitrogen compounds. *J. Gen. Microbiol.* **27**, 21–34.

Sørensen, L. H. (1975). The influence of clay on the rate of decay of amino acid metabolites synthesized in soils during decomposition of cellulose. *Soil Biol. Biochem.* **7**, 171–177.

Sorokin, Y. I. (1962). Experimental investigation of bacterial sulfate reduction in the Black Sea using ^{35}S. *Microbiology* **31**, 329–335.

Sorokin, Y. I. (1965). On the trophic role of chemosynthesis and bacterial biosynthesis in water bodies. *Mem. Ist. Ital. Idrobiol. Suppl.* **18**, 187–205.

Sorokin, Y. I. (1972). The bacterial population and the process of hydrogen sulfide oxidation in the Black Sea. *J. Cons. Intern. Explor. Mer.* **34**, 423–454.

Sorokin, Y. I. and Kadota, H. (1972). "Techniques for the assessment of microbial production and decomposition in fresh water". IBP Handbook No. 23. Blackwell, Oxford.

Srna, R. F. and Baggaley, A. (1975). Kinetic response of perturbed marine nitrification systems. *J. Water Poll. Con. Fed.* **47**, 472–486.

Stanford, G., Legg, J. D., Dzienia, S. and Simpson, E. C. Jr. (1975). Denitrification and associated nitrogen transformations in soils. *Soil Sci.* **120**, 147–152.

Stanford, G., Dzienia, S. and Van der Pol, R. A. (1975). Effect of temperature on denitrification in soils. *Soil Sci. Soc. Am. Proc.* **39**, 867–870.

Stanford, G. and Smith, S. J. (1972). Nitrogen mineralization potentials of soils. *Soil Sci. Soc. Amer. Proc.* **36**, 465–472.

Stanier, R. Y. (1974). The origin of photosynthesis in eukaryotes. *Soc. Gen. Microbial. Symp.* **24**, 219–240.

Stanier, R. Y., Douderoff, M. and Adelberg, E. A. (1970). "General Microbiology". 3rd edition. Macmillan, London.

Starkey, R. L. (1966). Oxidation and reduction of sulfur compounds in soils. *Soil Sci.* **101**, 297–306.

Stewart, W. D. P. (ed.) (1976). "Nitrogen Fixation by Free-Living Organisms". (IBPC). Cambridge University Press.

St. John, R. T. and Hollocher, T. C. (1977). Nitrogen 15 tracer studies on the pathway of denitrification in *Pseudomonas aeruginosa. J. Biol. Chem.* **252**, 212–218.

Stout, J. D., Tate, K. R. and Molloy, L. F. (1976). Decomposition processes in New Zealand soils with particular respect to rates and pathways of plant degradation. *In* "The Role of Terrestrial and Aquatic Organisms in Decomposition Processes" (J. M. Anderson and A. Macfadyen, eds.), pp. 97–144. Blackwell Scientific Publications, Oxford.

Stouthamer, A. H. (1973). A theoretical study of the amount of ATP required for microbial cell material. *Antonie van Leeuwenhoek* **39**, 545–565.

Stumm, W. and Morgan, J. J. (1970). "Aquatic Chemistry". John Wiley & Sons, New York.

Sturgis, M. B. (1936). Changes in oxidation-reduction equilibria in soils as related to the physical properties of the soil and the growth of rice. *La. Agr. Expt. Sta. Bull.* **271**.

Svensson, B. H. and Söderlund, R. (eds.) (1976). "Nitrogen, Phosphorus and Sulfur—Global Cycles". *Ecol. Bull.* **22**. Swedish Natural Science Council, Stockholm.

Swain, F. M. (1969). Palaeomicrobiology. *Ann. Rev. Microbiol.* **23**, 455–464.

Swain, F. M. (1970). "Non-Marine Organic Chemistry". Cambridge University Press, Cambridge.

Sybesma, C. (1969). Light-induced reactions of P890 and P800 in the purple photosynthetic bacterium *Rhodospirillum rubrum. Biochim. Biophys. Acta.* **172**, 177–179.

Sykes, R. M. (1975). Theoretical heterotrophic yields. *Jour. Water poll. Control Fed.* **47**, 591–600.

Tanaka, M. (1953). Occurrence of hydroxylamine in lake waters as an intermediate in the bacterial reduction of nitrate. *Nature* **171**, 1160–1161.

Taylor, B. F. and Heeb, M. J. (1972). The anaerobic degradation of aromatic compounds by a denitrifying bacterium. *Arch. Mikrobiol.* **83**, 165–171.

Taylor, D. L. (1970). Chloroplasts as symbiotic organelles. *Int. Rev. Cyt.* **27**, 29–64.

Taylor, D. L. (1973). The cellular interactions of algal-invertebrate symbiosis. *Adv. Mar. Biol.* **11**, 1–56.

Taylor, F. J. R. (1974). Implications and extensions of the serial endosymbiosis theory of the origin of eukaryotes. *Taxon.* **23**, 229–258.

Thauer, R. K., Jungermann, K. and Decker, K. (1977). Energy conservation in chemotrophic anaerobic bacteria. *Bact. Rev.* **41**, 100–180.

Thauer, R. K., Jungemann, K., Henninger, H., Wenning, J. and Decker, K. (1968). The energy metabolism of *Clostridium kluyveri. Europ. J. Biochem.* **4**, 173–180.

Thorstenson, D. C. (1970). Equilibrium distribution of small organic moecules in natural waters. *Geochim. Cosmochim. Acta.* **34**, 745–770.

Tirén, T., Thorin, J. and Nömmik, H. (1976). Denitrification measurements in lakes. *Acta. Agric. Scand.* **26**, 175–184.

Tredway, J. V. and Burton, S. D. (1974). Morphological examination of *Beggiatoa* and *Thiothrix* obtained from bacterial mats on the surface of solid waste bales deposited in the continental shelf. *Bact. Proc. G* 105.

Tribe, H. T. (1961). Microbiology of cellulose decomposition in soil. *Soil Sci.* **92**, 61–77.

Trimble, R. B. and Ehrlich, H. L. (1968). Bacteriology of manganese nodules. III. Induction of MnO_2 by two strains of nodule bacteria. *Appl. Microbiol.* **16**, 695–702.

Trimble, R. B. and Ehrlich, H. L. (1970). Bacteriology of manganese nodules. IV. Reduction of an MnO_2-reductase system in a marine bacillus. *Appl. Microbiol.* **19**, 966–972.

Trudinger, P. A. (1967). Metabolism of thiosulfate and tetrathionate by heterotrophic bacteria from soil. *J. Bacteriol.* **93**, 550–559.

Trudinger, P. A., Lambert, I. B. and Skyning, G. W. (1972). Biogenic sulfide ores: a feasibility study. *Econ. Geol.* **67**, 1114–1127.

Tuffey, T. J., Hunter, J. V. and Matulewich, W. A. (1974). Zones of nitrification. *Bull. Amer. Water Res. Assoc.* **10**, 1–10.

Tuovinen, O. H., Nicomelä, S. I. and Gyllenberg, H. G. (1971). Tolerance of *Thiobacillus ferrooxidans* to some metals. *Antonie van Leeuwenhoek* **37**, 489–496.

Tuttle, J. H. and Jannasch, H. W. (1973). Sulfide and thiosulfate oxidizing bacteria in anoxic marine basins. *Marine Biology* **20**, 64–70.

Tyler, P. A. (1970). Hyphomicrobia and the oxidation of manganese in aquatic systems. *Antonie van Leeuwenhoek* **36**, 567–578.

Uzzell, T. and Spolsky, C. (1974). Mitochondria and plastids as endosymbionts: a revival of special creation? *Amer. Sci.* **62**, 334–343.

Vaccaro, R. F., Hicks, S. E., Jannasch, H. W. and Currey, F. G. (1968). The occurrence and role of glucose in sea water. *Limnol. Oceanogr.* **13**, 356–360.

Vanderborght, J.-P. and Billen, G. (1975). Vertical distribution of nitrate concentration in interstitial water of marine sediments with nitrification and denitrification. *Limnol. Oceanogr.* **20**, 953–961.

Vanderborght, J. P., Wollast, R. and Billen, G. (1977). Kinetic models of diagenesis in disturbed sediments. Part 2. Nitrogen diagenesis. *Limnol. Oceanogr.* **22**, 794–803.

Van der Linden, A. C. and Thijsse, G. S. E. (1965). The mechanism of microbial oxidations of petroleum hydrocarbons. *Adv. Enzymol.* **27**, 469–546.

Van Gemerden, H. (1967). "On the bacterial sulfur cycle of inland waters". Ph.D. thesis, Leiden, 110 pp.

Van Gent-Ruijters, M. L. W., de Vries, W. and Stouthamer, A. H. (1975). Influence of nitrate on fermentation pattern, molar growth yield, and synthesis of cytochrome *b* in *Propionibacterium pentosaceum*. *J. Gen. Microbiol.* **88**, 36–48.

Vagnai, S. and Klein, D. A. (1974). A study of nitrite-dependent dissimilatory microorganisms isolated from Oregon soils. *Soil Biol. Biochem.* **6**, 335–339.

Van Kessel, J. F. (1977). Factors affecting the denitrification rate in two water-sediment systems. *Water Res.* **4**, 259–267.

Van Ravenswaay, C. J. C. and Van der Linden, A. C. (1971). Substrate specificity of the paraffin hydroxylase of *Pseudomonas aeruginosa*. *Antonie van Leeuwenhoek* **37**, 339–352.

Van Veen, W. L. (1972). Factors affecting the oxidation of manganese by *Sphaerotilus discophorus*. *Antonie van Leeuwenhoek* **38**, 623–626.

Veldkamp, H. and Jannasch, H. W. (1972). Mixed culture studies with the chemostat. *J. Appl. Chem. Biotechnol.* **22**, 105–123.

Verstraete, W. and Alexander, M. (1972). Heterotrophic nitrification in samples from natural environments. *Naturwissenschaften* **59**, 79–80.

Vincent, J. M. (1974). Root-nodule symbiosis with *Rhizobium. In* "The Biology of Nitrogen Fixation" (A. Quispel, ed.), pp. 265–341. North Holland Publishing Company, Amsterdam.

Vosjan, J. H. (1975). Ecological and physiological aspects of bacterial sulfate reduction in the Wadden Sea. (In Dutch.) Thesis, Groningen, 68 pp.

Wada, E., Kadonga, T. and Matsuo, S. (1975). [15]N abundance in nitrogen of naturally occurring substances and global assessment of denitrification from isotopic viewpoint. *Geochem. J.* **9**, 139–148.

Wakao, N. and Furusaka, C. (1976). Presence of micro-aggregated containing bacteria in a paddy field soil. *Soil Biol. Biochem.* **8**, 157–159.

Walker, D. J. (1968). The position of lactic acid and its derivatives in the nutrition and metabolism of ruminants. *Nut. Abs. and Rev.* **38**, 1–11.

Walsh, F. and Mitchell, R. (1972). A pH-dependent succession of iron bacteria. *Environ. Sci. Tech.* **6**, 809–812.

Walter, M. R., Bauld, J. and Brock, T. D. (1972). Silicaceous algal and bacterial stromatolites in hot spring and geyser effluents of Yellowstone National Park, *Science* **178**, 402–405.

Weiler, R. R. (1973). The interstitial water composition in the sediments of the Great Lakes. I. Western Lake, Ontario. *Limnol. Oceanogr.* **18**, 918–931.

Wertleib, D. and Visniac, W. (1967). Methane utilization by a strain of *Rhodopseudomonas gelatinosa. J. Bacteriol.* **93**, 1722–1724.

Wetzel, R. G., Rich, P. H., Miller, M. C. and Allen, H. L. (1972). Metabolism of disdolved and particulate detrital carbon in a temperature hard-water lake. *Mem. 1st. Ital. Idrobiol. Suppl.* **29**, 185–243.

Whittenbury, R., Dalton, H., Eccleston, M. and Reed, H. L. (1975). The different types of methane oxidizing bacteria and some of their properties. *In* "Microbial Growth on C$_1$ Compounds", pp. 1–9. The Society of Fermentation Technology, Tokyo.

Whittenbury, R. and Kelly, D. P. (1977). Autotrophy: a conceptual Phoenix. *Symposia Soc. Gen. Microbiol.* **24**, 121–149.

Whittenbury, R., Phillips, K. C. and Wilkinson, J. F. (1970). Enrichment, isolation and some properties of methane-utilizing bacteria. *J. Gen. Microbiol.* **61**, 205–218.

Whitton, B. A. (1971). Terrestrial and freshwater algae of Aldabra. *Phil. Trans. Roy. Soc. B.* **260**, 249–255.

Whitton, B. A. (1973). Freshwater plankton. *In* "The Biology of Blue-Green Algae" (N. G. Carr and B. A. Whitton, eds.), pp. 353–367. Blackwell Scientific Publications, Oxford.

Whitton, B. A. and Sinclair, C. (1975). Ecology of blue-green algae. *Sci. Prog. Oxford* **62**, 429–446.

Widdel, F. and Pfennig, N. (1977). A new anaerobic, sporing, acetate-oxidizing, sulfate-reducing bacterium, *Desulfotomaculum* (emend.) *acetoxidans. Arch. Microbiol.* **112**, 119–122.

Wiegert, R. G. and Owen, D. F. (1971). Trophic structure, available reserves and population density in terrestrial vs. aquatic ecosystems. *J. Theor. Biol.* **30**, 69–81.

Wieringa, K. T. (1940). The formation of acetic acid from carbon dioxide and hydrogen by anaerobic sporeforming bacteria. *Antonie van Leeuwenhoek.* **6**, 251–262.

Wijler, J. and Delwiche, C. C. (1954). Investigations on the denitrifying process in soils. *Plant and Soil* **5**, 155–169.

Williams, P. J. le B. (1970). Heterotrophic utilization of dissolved organic compounds in the sea. I. Size distribution of population and relationship between respiration and incorporation of growth substrates. *J. Mar. Biol. Assoc. U.K.* **50**, 859–870.

Williams, P. J. le B. (1973). The validity of the application of a simple kinetic analysis to heterogeneous microbial populations. *Limnol. Oceanogr.* **18**, 159–165.

Winfrey, M. R. and Zeikus, J. G. (1977). The effect of sulfate on carbon and electron flow during microbial methanogenesis in fresh water sediments. *Appl. Environ. Microbiol.* **33**, 275–281.

Winogradsky, H. (1949). Contribution to the study of nitrifying microflora of sewage: resistance of the bacteria to unfavourable conditions. *Annls. Inst. Pasteur, Paris,* **76**, 35–42.

Winogradsky, S. and Winogradsky, H. (1933). Etudes sur la microbiologie du sol. Nouvelles recherches sur les organismes de la nitrification. *Annls. Inst. Pasteur, Paris,* **50**, 350–432.

Witkamp, M. and Ausmus, B. S. (1976). Processes in decomposition and nutrient transfer in forest systems. *In* "The Role of Terrestrial and Aquatic Organisms in Decomposition Processes" (J. M. Anderson and A. Macfadyen, eds.), pp. 375–396. Blackwell Scientific Publications, Oxford.

Woese, C. R., Pribula, C. D., Fox, G. E. and Zablen, L. B. (1975). The nucleotide sequence of the 5S ribosomal RNA from a photobacterium. *J. Mol. Evol.* **5**, 35–46.

Wojtalik, T. A. (1970). Elements of a program for carbon–oxygen budgets in reservoirs. *In* "TVA Activities Related to Study and Control of Eutrofication in the Tennessee Valley", pp. 29–33. National Fertilizer Development Center, Muscle Shoals, Ala.

Wolfe, R. S. (1971). Microbial formation of methane. *Adv. Microbial Physiol.* **6**, 107–146.

Wolfe, R. S. and Pfennig, N. (1977). Reduction of sulfur by *Spirillum* 5175 and syntrophism with *Chlorobium. Appl. Environ. Microbiol.* **33**, 427–433.

Wolin, M. J. (1974). Metabolic interactions among microorganisms. *Ann. J. Clin. Nutr.* **27**, 1320–1328.

Wood, E. J. F., Odum, W. E. and Zieman, J. C. (1969). Influence of sea grasses on the productivity of coastal lagoons. *In* "Lagunas Costeras, un Simposio", pp. 495–502. Mem. Simp. Intern. Lagunas Costeras. UNAM-UNESCO, Mexico.

Wood, J. M. (1974). Biological cycles of toxic elements in the environment. *Science* **183**, 1049–1051.

Wood, L. W. (1973). Monosaccharide and disaccharide interactions on uptake and catabolism of carbohydrates by mixed microbial communities. *In* "Estuarine Microbial Ecology" (L. H. Stevenson and R. R. Colwell, eds.), pp. 181–197. University of South Carolina Press, Columbia, S.C.

Woodwell, G. (1970). The energy cycle of the biosphere. *Sci. Amer.* **223**:3, 64–74.

Woolfolk, C. A. and Whiteley, H. R. (1962). Reduction of inorganic compounds by *Micrococcus lactilyticus. J. Bacteriol.* **84**, 647–658.

Wright, R. T. (1973). Some difficulties in using ^{14}C-organic solutes to measure heterotrophic bacterial activity. *In* "Estuarine Microbial Ecology" (L. H. Stevenson and R. R. Colwell, eds.), pp. 199–217. University of South Carolina Press, Columbia, S.C.

Wright, R. T. and Hobbie, J. E. (1965). Uptake of organic solutes in lake water. *Limnol. Oceanogr.* **10**, 22–28.

Wright, R. T. and Hobbie, J. E. (1966). Use of glucose and acetate by bacteria and algae in aquatic ecosystems. *Ecology* **47**, 447–464.

Wright, R. T. and Shah, N. M. (1975). The trophic role of glycollic acid in coastal seawater. I. Heterotrophic metabolism in seawater and bacterial cultures. *Mar. Biol.* **33**, 175–183.

Yamanaka, T. (1964). Identity of *Pseudomonas* cytochrome oxidase with *Pseudomonas* nitrite reductase. *Nature,* 253–255.

Yčas, J. W. (1976). The palhirrotrophic origin of energy metabolism. *Origins of Life* **7,** 235–238.

Yoneyama, T. and Yoshida, T. (1977). Decomposition of rice residue in tropical soils. *Soil Sci. Plant Nut.* **23,** 33–48.

Yoshinari, T., Hynes, R. and Knowles, R. (1977). Acetylene inhibition of nitrous oxide reduction and measurement of denitrification and nitrogen fixation in soil. *Soil Biol. Biochem.* **9,** 177–183.

Zajic, J. A. (1969). "Microbial Biogeochemistry". Academic Press, New York and London.

Zavarzin, G. A. (1964). *Metallogenium symbioticum. Ztschr. Allg. Mikrobiol.* **4,** 390–395.

Zavarzin, G. A. (1968). Bacteria in relation to manganese metabolism. *In* "The Ecology of Soil Bacteria" (T. R. G. Gray and D. Parkinson, eds.), pp. 612–623. University of Toronto Press, Toronto.

Zeikus, J. G., Weimer, P. J., Nelson, D. R. and Daniels, L. (1975). Bacterial methanogenesis: acetate as a precursor in pure culture. *Arch. Microbiol.* **104,** 129–134.

Zeikus, J. G. and Winfrey, M. R. (1976). Temperature limitation of methanogenesis in aquatic sediments. *Appl. Environ. Microbiol.* **31,** 99–107.

ZoBell, C. E. (1943). The effect of solid surfaces upon bacterial activity. *J. Bacteriol.* **46,** 39–56.

ZoBell, C. E. (1964). The occurrence, effects and fate of oil polluting the sea. *Internatl. Confr. Water Poll. Res. London* 1962, pp. 85–118, Pergamon Press, London.

ZoBell, C. E. and Upham, H. C. (1944). A list of marine bacteria including descriptions of sixty new species. *Bull. Scripps. Inst. Oceanog.* **5,** 239–292.

Index

213

Z